A Celebration of the Static Scale Airplane Modeller's Craft

SCRATCH BUILT!

John Alcorn, George Lee, & Peter Cooke

(Dedicated to Paul Matt)

Schiffer Military/Aviation History
Atglen, PA

On the front cover:
George Lee's Keystone B4A

On the rear cover:
Top: John Alcorn's Laird Super Solution
Bottom: Peter Cooke's Spitfire Mk.IX

Book Design by Robert Biondi

First Edition
Copyright © 1993 by John Alcorn, George Lee, Peter Cooke.
Library of Congress Catalog Number: 92-62607

All rights reserved. No part of this work may be reproduced or used in any forms or by any means – graphic, electronic or mechanical, including photocopying or information storage and retrieval systems – without written permission from the copyright holder.

Printed in the United States of America.
ISBN: 0-88740-417-0
We are interested in hearing from authors with book ideas on related topics.

Published by Schiffer Publishing Ltd.
77 Lower Valley Road
Atglen, PA 19310
Please write for a free catalog.
This book may be purchased from the publisher.
Please include $2.95 postage.
Try your bookstore first.

Contents

ACKNOWLEDGEMENTS 6
In Memoriam: George Lee

INTRODUCTION 10
Health Warning

CHAPTER I: ROOTS: THE ORIGINS OF SCRATCH BUILDING STATIC SCALE MODEL AIRCRAFT 17

- The Antebellum Era
- Wartime Development
- The Righteous Solid Model
- The Plastic Era
- IPMS

CHAPTER II: HARD CORE LORE 28

- Commercial Plans Sources
- Archival Plans/Configuration Sources
- Background Material
- Scratch Built Plans: Some Case Histories
- Modelling Techniques

CHAPTER III: VACUFORMING 45

- Carving the Forms
- Mounting the Forms
- The Vacuforming Rig
- Vacuforming Technique
- Canopies

CHAPTER IV: THE BASIC MODEL STRUCTURE 51

- Fuselage:
  - Vacuformed Shells
  - Slab Sides

- Wings:
  - Solid
  - Shell Over Structure
  - Rib, Spar and Skin Construction

- Surface Features

CHAPTER V: RESIN CASTING 67

- Introduction
- Mold and Casting Materials
- Major Components
- Engines, Cowlings and Propellers
- Painting

CHAPTER VI: DETAILS 81

- Cockpits/Visible Compartments
- Guns
- Landing Gear
- Engines and Propellers

CHAPTER VII: ASSEMBLY 92

- Non-Rigged A/C
- Rigged A/C

CHAPTER VIII: PAINTING 100

- Airbrushing
- Weathering and Scale Effect
- Surface Effects
- Colors

CHAPTER IX: MARKINGS 113

- Silk Screening
- Air Brushed Markings
- Hand Painting

EPILOGUE 121

POSTSCRIPT: 122
Model Portrait Photography

THE MODELLER'S LAMENT

O' The angular line of a 109,
The elegant grace of a Spit.
The bridgebuilder's maze of a Brandenburg,
The Stirling's outrageous sit.

O' To bash a kit of a Messerschmitt,
Is a childhood dream come true.
A year or so on a scratchbuilt job,
It takes that long to do.

With devotion to labor on a Stuka or Sabre,
Shaping cannons, and bombs that fall.
Can I ever explain that I'm really quite sane,
And not a warmonger at all?

Bucket seats, rigging cleats, longerons and stringers,
Struts cabane and interplane, a microscopic lever.
Townend rings and little things,
To pop in the air and into your hair
And disappear forever.

A surgeons' hand, a jewelers eye,
To show each bloody rivet.
An artists' skill with knife or drill,
Can you replant an epidermal divot?

Others may tingle with lust as they jingle,
The lucre they've strived to accrue.
I'd surely prefer to such weakness defer,
And spend all of my lucre on glue.

Still others may moan, it's not silver or stone,
But plastic and therefore a toy.
Venus' very own daughter was 68% water,
Her form, not her substance brought joy.

Some I could mention regard a convention
As an excuse for boozin' and sinnin'.
But at IPMS, the stress we confess
Is all on the sinnin' of winnin'.

– John Alcorn

ACKNOWLEDGEMENTS

Twenty years ago, while colleagues at the Stanford Linear Accelerator Center (SLAC), George Lee and I vowed to do a book on scratch building. But, as with so many fantasies, reality and the crush of events intervened. (Just as well, since the scratch-built saga is a great deal richer by now.) Then, in late 1988, during a business trip to the S.F. Bay Area from my new home in Williamsburg, I visited George and beheld his Keystone B4A, which had recently Mark Spitzed the IPMS/USA Nationals in Dayton. I exclaimed, "George, that does it! We must begin the book," adding the cautionary note that "we're not exactly kids anymore" (physically, at least). I began writing on the flight home. Then, during a Battle of Britain 50th pilgrimage to England, I visited Peter Cooke at his home/workshop in Sonning-on-Thames. His models are paragons of workmanship and realistic appearance; and his resin cast technique represents an attractive alternative to vacuforming. So, we invited him as a co-author.

George's Keystone set the spark which brought this book to light. The core of modelling lore residing between these pages was provided by us, as well as many model portrait photos: George Lee prepared all of the line drawing illustrations. But, Scratch Built! would never have been without the generosity of Bob Rice, Bill Bosworth, Arlo Schroeder, Bob Davies, Alan Clark, Ron Lowry, Paul Budzik, Ron Cole, and Harry Woodman, whose model portraits grace these pages. To them, we extend our deepest gratitude.

Robert Mikesh has given much needed moral support, as well as valuable assistance in locating and contributing photos of models now in the NASM. Those of us who have built models for NASM exhibits did so largely due to our respect for, and the persuasion of Bob, who convinced us that "psychic satisfaction" was ample reward for our efforts. We also appreciate the support and assistance rendered by Phil Edwards of the NASM staff.

Credit for the Harmonic Convergence between *Scratch Built!* and Schiffer Publishing is due to John "Hawkeye" Campbell, who spotted my "Hello, My Name Is . . ." badge across a crowded vendor room at the 1991 IPMS/USA Convention in St. Louis; made the connection with *The Jolly Rogers: The 90th Bomb Group in WWII*; and thereupon introduced me to Peter Schiffer of Hopi Kachina Dolls to The 12th SS Panzer Division publishing fame. A month later we were under contract.

Schiffer Publishing, Ltd. of Atglen, PA, is a wonderful anachronism: a very successful business, employing state-of-the-art techniques/equipment in a bucolic turn-of-the-century setting, staffed by an extended family of cheerful employees. The main operation is housed in a huge 1902 vintage barn amid a typical prosperous Pennsylvania farm complex, replete with horses and Black Angus cows. Despite appearances, they produced 90 titles in 1991 and have some 600 in print! Scratch Built!s youngish editor, Bob Biondi, came to this calling from seven years as a chef; Peter Schiffer entered publishing via the antique furniture restoration biz; and his

Acknowledgements

This fine 1/18th scale model of the NASM's exquisitely restored Albatros DVa was built by Alan Clark of England. It graced the cover of the May/June 1990 Windsock, which featured a copiously illustrated article on its construction. Alan's inspiration for the project was Robert Mikesh's Smithsonian book describing restoration of STROPP. Needless to say, he worked from Bob Waugh's plans. The fuselage halves were wet-lacquered fiberglass over pine forms. After assembly of the two halves, the exterior surface was wood veneered panel-by-panel, following the pattern of the actual aircraft.

---

wife, Nancy, had worked for the Smithsonian. Ambiance.

Typing – "pixel pecking" of the *Scratch Built!* manuscript was performed admirably by Leigh Ann Garza, under a "time and materials" agreement. She not only deciphered my ever-deteriorating penmanship (akin to breaking the Enigma code), but coped unflinchingly with terms like Schwarzlose, rigging jigging, Kookaburra and polystyrene stick stock.

Authors are always careful to acknowledge the forbearance of their spouses. In my case however, *Scratch Built!* is simply the latest of a dreary litany of indignities endured by my wife over the past 30-some odd years, ranging from old car restoration, through the *Life* magazine, phonograph/record and military helmet collections, to ad nauseam recitation of Chaucer's Prologue to the Canterbury Tales – to say nothing of malodorous model mess.

As with so many deeply rewarding activities, modelling demands long periods of intense, solitary concentration, punctuated by occasional displays of manic gregariousness (IPMS meetings and conventions). There are other, less emotive interactions with modelling/aviation historian/delineator colleagues, during which we eagerly share lore of mutual interest. So, each of us owes a great debt to others for inspiration, learning and assistance during the course of our modelling careers.

Modellers whose fellowship we value include Don Alberts, Bob LaBouy, Phil Huston, Walt Fink, Ed Boll, Tom Walsh, Pat Stein, John Ficklin, Chris Mikuriya, Pete Chalmers, Wayne Wachsmuth, Tom Darcy and others too numerous to mention – in addition to those specifically named in the text. In my case, I must also mention Alvis (later Jarmin) Lynch, with whom I shared my early modelling years in Houston. In their various ways, these folk have also provided inspiration for Scratch Built! We hope that they find it entertaining, and perhaps even useful.

In Memoriam: George Lee

George's passing came as a great shock and personal loss to me, since we have been friends for over twenty years.

All friendships are based upon some common bond, such as having been childhood playmates, school chums, service buddies, professional colleagues, or sharing a common avocational interest. George and I worked together at SLAC and shared an abiding interest in vintage airplanes and model building.

Professionally, George was a top notch mechanical designer, who made many contributions to physics research during his long, productive career at Livermore and SLAC.

As a builder of static scale model aircraft, George was simply the best of all. This was due to several factors: his vast knowledge of vintage aircraft and aviation history; his superb skill and boundless patience; his resourcefulness at developing techniques to solve challenging problems; and eating rice, as he often reminded me.

Unrealized by most of his modelling associates was his second hobby as a sculptor, in wood, of marine life.

But even beyond these talents was George the person: his warmth, humor, consideration, judgement and integrity. Evidence of these attributes are his many friends, devoted wife Milly and fine children Michael, Terry and Beverly. I have never known anyone for whom I have had greater respect.

While I will miss George deeply, I'm grateful for our friendship and pleased that we were able to accomplish our long cherished desire to produce a book on scratch building model airplanes. I hope that it will serve as a worthy testament to his modelling career.

John Alcorn
6 June 1992

NOTE: At the 1992 IPMS/USA National Convention in Seattle, it was announced that henceforth the Judge's Grand Award will be called the George Lee Award.

(Ben Walker photo)

In Memoriam: George Lee 9

INTRODUCTION

Welcome aboard, fellow modeller! Within the confines of this book we will take you on a tour of the scratch builder's domain. By this we mean the history, infrastructure, technique and examples of "one-off" static scale model aircraft, constructed from basic raw materials; primarily high impact polystyrene plastic sheet.

First, we will subject you to a heavy dose of modelling nostalgia, leading you through development of the hobby from its solid (wood) model origins. Along the way, we'll digress somewhat to trace the emergence of plastic kits. After all, we are kindred spirits: many modellers are AC/DC, as they say today. Besides, there is a broad grey realm where the kit superdetailer/conversionist blends imperceptibly into the scratch builder, who shamelessly pilfers engines, propellers, and such from kits.

We'll explore the store of hard core lore which constitutes the foundation of our historical aviation interest, as well as providing the bulk of our project source material. Here again, we'll wax nostalgic, leading you down musty, dimly lit halls of yellowing, thumbstained archives which fueled the fires of our youthful enthusiasm. But, lest we lose you in the stacks, we won't tarry long. Soon enough, we'll emerge into the sunlit uplands of modern grist, where we'll survey the wealth of fine source material available today.

BELLANCA "CRUISAIR": This spectacular model by Bob Rice represents one of Bellanca's best known products, featuring their trademark lifting wing struts. At the 1987 IPMS/USA Nationals in Washington, DC, it shared scratch-built honors with Ron Lowry's Cruisair! – the latter on floats. (Rice photo)

Introduction 11

CURTISS "CARRIER PIGEON": Some 36 years after having seen one at Mills Field, San Francisco and 48 years after its "heyday," George Lee immortalized this "clunker" in polystyrene, working from Paul Matt drawings. Ten Liberty engined Carrier Pigeons were built for NAT in 1926 to fly the mail. (Joe Faust photo)

Then, having sampled these preprandial delights, we'll delve into the main course: the meat, potatoes and condiments of scratch building technique. While thus gorging you on such food for thought, we intend to satiate your visual appetite with a profusion of photos showing voluptuous models in revealing full color poses.

OUR OWN ROOTS: Every old hand was once a novice. Each of us has come to our hobby by some unique process of discovery, inspiration and learning. So, please indulge us as we briefly relate our own stories.

John Alcorn: I made my appearance on 29 February 1932, the night that Charles and Anne Morrow Lindbergh's baby was kidnapped, and two days after Elizabeth Taylor's debut.

While I became fascinated by airplanes very early, my first coherent memory of model building was in 1940 when, at age 8, I constructed a Curtiss P-40 from a HAWK kit.

The event which kindled my youthful aviation enthusiasm more than any other was a 1941 Bundles for Britain exhibit (in Houston) of an "Me 109", shot down during the Battle of Britain. Sporting a yellow nose, tiger insignia and 5 "kill" markings on the tail, it was an awesome sight to an airplane crazy kid of nine as I peered into the open cockpit from the wood scaffolding over the wing root. (NOTE: As we all know now, it was Werke Nummer 1190, a Bf109 E-3, "Weiss 4" of 4/JG 26, flown by Uffz. Horst Perez when brought down on 30 September 1940).

During the War years, I ran a veritable production line of solid models, initially from the 1/72nd scale recognition model plans. Models continued to emanate from my "workshop" until 1950 when college presented other diversions. This was followed by two years in the Navy, during which time I produced my only jet ever: an F7U Cutlass, fashioned on the wardroom table of DD885.

From 1959 through 1972 I averaged one solid model per year, mostly WWII types to larger scales.

In about 1969, I had just completed a 1/24th scale "solid" Bf-109E when I made the acquaintance of George Lee, a colleague at the Stanford Linear Accelerator Center (SLAC). As he examined my 109, I could tell that he felt I needed professional help (Chinese-Americans aren't necessarily inscrutable). Oh, the carving was fine and, though hand painted, it was neat enough. Following construction of a 1/24th scale Hurricane MKI, a second Bf-109E-4 sported an airbrushed, mottled camouflage scheme and full 9./JG 2 heraldry from

SCRATCH BUILT!

So, here we have your junior author (who was a mere 10-1/2 at the time of this photo), shortly after the 1974 IPMS/USA Nationals. Haircuts are not high on the list of modeller priorities. (Joe Faust photo)

home made decals, thanks to George's tutelage. There was no turning back.

By this time, George was converting to vacuformed plastic, whereupon he convinced me to follow suit. Thus was born my 1/32nd scale Douglas A-20A, described elsewhere. This was followed by a Laird Super Solution for the NASM, a Rumpler CIV, and a Wedell-Williams #44 for the NASM, plus one for me. Following a four year hiatus to restore two old cars (a 55 Chevy and 50 Olds), I returned to the fold to write this book and construct a D.H.9A in 1920s livery.

George Lee: I entered this world on March 4, 1923 in San Francisco, born of immigrants from China. My father was FOB (Fresh Off the Boat) from Canton, having arrived aboard the S.S. Siberia in 1909.

I have been interested in airplanes ever since I can remember. The first model I ever built was a rubber-powered balsa wood kit which I received as a Christmas present from my Sunday School Teacher when I was 8 or 9 years old. It didn't fly!

When I was around 11 years old and in the sixth grade, I joined a craft club at the Y.M.C.A., where I started to develop my basic modelling and wood carving skills. Entering the flying and solid model contests and winning some awards were nice, but I really enjoyed the kits, tools, and magazines that I had received as door prizes or for participating in some of the contests.

In the late 1930s my interest in airplanes was greatly increased by aviation activities in the San Francisco Bay area. I remember large formations of Army and Navy aircraft flying over the city on the various holidays, the Akron and Macon flyovers, the China Clipper taking off for the Orient, the fleet of warships entering San Francisco Bay, the Sikorsky S-39s flying back and forth to the 1939 San Francisco World's Fair at Treasure Island, and visits to the many airfields around the Bay area. I was at Mills Field, now San Francisco International Airport, where I saw a Curtiss Carrier Pigeon. My first impression was "what a clunker!" At 14, I had my first airplane ride, in a Fleet trainer, at San Francisco Bay Airdrome in Alameda. I was given the flight after helping the pilot wipe the oil and dirt off the lower surfaces of another airplane. These events greatly influenced the choice of models for me to build.

I served in the U.S. Navy during WWII as an aircraft instrument technician – worked on almost every type of Navy aircraft used during the War.

I specialized in scratchbuilding 1/72 WWI models, using basswood, before the appearance of plastic models. I joined

Just to prove that we've been at it for awhile, here is your coauthor at age 13 (summer 1945), displaying his "solid" P-51D. The oval belt pouch is for Coin of the Realm – doubtless soon to be squandered at the local hobby shop (in "The Village", S/W Houston). (Alcorn photo)

Introduction

Here we have George proudly holding his Curtiss Carrier Pigeon. (Joe Faust photo)

the IPMS in the early 1960s and started scratch building with plastic material in 1969.

I won the Best of Show trophy at the 1971 IPMS National Convention held in Atlanta, Georgia, with a 1/32 scale Aviatik Berg D1., and served as Chairman of the 1977 IPMS National Convention at San Francisco.

I built a model of the Verville Sperry R.3 Racer, winner of the 1924 Pulitzer trophy race and a model of the Sikorsky S-39, Osa and Martin Johnson's airplane used in exploring Africa and Borneo, for the Smithsonian's National Air and Space Museum (NASM).

Then, I built a model of the "Rosamonde", the first airplane manufactured in China, for the Taiwan National Air Museum.

At the 1988 IPMS National Convention at Dayton, Ohio, I entered a Keystone Bomber (B-4A) and won the Best in Class (scratch-built), Judge's Best Aircraft, Detail and Scale's best detailed aircraft, Judge's Grand Award, and the most popular model (Best of Show), the most awards ever given to a single model in IPMS national model contest history. (George passed away on 31 May 1992).

Peter Cooke: I feel such a youngster alongside a pair of such hoary old gents (C'mon, Peter, we recognize this as an euphemism for "goats"). I missed the Battle of Britain completely, but I was born on the night of the great Blitz on London, 29th December 1940. Thirty-seven miles away in Henley-on-Thames we had a pretty quiet war: only three bombs, jettisoned by fleeing aircraft. Total casualties: one cow. Brought up in a country that was like a giant aircraft carrier, it is hardly surprising that I started modelling aircraft quite early. I remember the Penguin kits post-war, and I also built a few rubber-powered flying models. However, I lost interest in aircraft when the jets came in, turning instead to railways and cars.

When my son was six years old, in 1973, he was given a 1/72nd inch Airfix Spitfire IX, which I built for him. Thus, I discovered plastic kits! I built those for two years, but then felt the need for something more challenging. I also wanted to build larger and more detailed models. I started by super-detailing the Airfix 1/24th scale Hurricane, using the Airfix/PSL book on the subject. At this time I was awestruck by the beautiful sight of the exposed Griffon engine on the Spit 24 at the Hendon Museum. I converted the Airfix 1/24th scale Spit I into a Mk XIV with a completely scratch-built nose with detailed Griffon installation.

Here is a 1/24th scale wood model displaying the first fruits of my association with George Lee: silk screened decals and airbrushed paint scheme. The markings are for "Gelb 4" of 9./JG 2 "Richthofen", operational with Luftflotte 3 during the Battle of Britain. The configuration basis for the model was MAP Plan Pack BH2790 by Doug Carrick. (Alcorn photo)

PETER COOKE & ARTIFACT: Here is our token Redcoat, looking rather smug at having completed another Lancaster. (Cooke photo)

SCRATCH BUILT!

SPITFIRE Mk.IX: This shot reveals Cooke's panel, Dzus fitting and other detail to good effect. It would be difficult to tell this from a comparable photo of the real thing. (Cooke photo)

I then moved back to the Reading area, where I still live, and joined a prominent IPMS branch. They persuaded me to enter my Spitfire in the forthcoming National Championships, and it won the John Edwards Trophy. A fellow member at Reading was a hero of mine, Tony Woollett, whose models I had admired at the Model Engineers Exhibition, and who was a great source of inspiration. The following year, 1977, I created my first completely scratch-built model, a Hawker Tempest Mk V. It won "Best of Show" at the IPMS Nationals that year. The following year I scratch-built a Hawker Sea Fury which won the Championship Cup at the International Model Engineers Exhibition. Soon after this, following sixteen years of teaching Physics, I became a professional modelmaker, specializing in 1/24th scale aircraft models. By casting components in epoxy resin, using silicone rubber moulds, I am able to spread the considerable costs of research and pattern-making over a batch of models.

OUR DUBIOUS MOTIVES: We must ask ourselves: why would an otherwise reasonably normal person forsake the pleasures of family, garden, TV, bowling, hunting, fishing, sailing, barhopping, vacationing, fixing faucets and playing poker to sit for years at a time, hunched over a cluttered workbench, squinting through an Optivisor, fashioning tiny bits of white plastic into what eventually emerges as a miniature representation of a long-extinct piece of machinery which few still recall and about which fewer give a damn?

Perhaps the motivation is the noble urge to excel: to create a unique masterpiece from a few dollars worth of scrap material; to serve the cause of history by faithfully preserving the appearance of some ancient engine of war or commerce; or to vicariously relive the romance, the courage, the endurance, the thrill and the mortal terror of the dramatic events in which the actual machines participated.

The answer, of course, is none of the above: but rather to take the completed object to a large gathering of similarly inclined colleagues, and to return with a largish pseudo-neo Gothic likeness of the Tower of Babel, fashioned in metallized plastic.

We may be a bit queer, but we're not the worst. Some years ago, Bob Meuser took me by the home of Earl Thompson in Livermore, California to see his world-class R/C flying scale models. It was a scene of carnage: as I recall, the only model still intact was a Swedish FW-44J Stieglitz biplane which was nearing completion. A Westland Lysander of epic proportions and awesome detail reposed forlornly on the dining room table, its landing gear collapsed and one wing crumpled back along the fuselage. It only got worse as we continued the tour. Yet, he was quite cheerful about the whole thing, explaining that it's all part of the hobby, and besides, they're all repairable! Different strokes for different folks.

In this tome, we're not going to tread on others' turf, since we'd probably just bog down. We don't do zimmerit, camp followers, English gardens, poop-decks, nerf-bars, rumble-seats or cowcatchers. But that doesn't mean that we don't occasionally fraternize with some of those nut cases. For example, that nervous looking guy in the raincoat surreptitiously eyeing the photo-etched parts in the Grey Whale Ship Shop just might be one of our own kind, seeking goodies for his SBC-4 cockpit interior.

Introduction

DETAIL OF DH9A: No, this is not a shot of the RAF Museum's F1010, but a 1/6th scale radio controlled model by Peter McDermott of England. It's difficult to imagine the concept of deliberately launching a work of art such as this into the ether. Needless to say, this incredible model won the International Scale class at the 1988 British RC Nationals, and scored top static points at the RC World Championship, held that year in Italy. (McDermott photo)

BUT SERIOUSLY . . . We believe that, as a general rule, each succeeding model should be your personal best – your *magnum opus*. Perhaps it's a more complex subject, has a more challenging paint scheme, or requires some technique which you have not previously attempted – such as scratch building. But, always you (we) should strive to improve upon workmanship and realism – Peter Cooke will expand upon the latter topic in Chapter V.

Earlier, we gently chided ourselves about modelling to win the Gold. We'd be foolish to suggest that it's not important. Peer approval is certainly important to us all, and competition is a great stimulus to achievement.

But, the most rewarding goal is psychic satisfaction: knowledge that the finished product really is good – reflecting the best you had within you. Sure, it was an epic struggle, but it came out well, and now you can rest upon your laurels for awhile. You go out and lie in the hammock all afternoon (or until the bugs find you); rent a few old movies (Test Pilot, Men with Wings, Dawn Patrol – the usual fare); go to the zoo. But, before long, that old urge begins to return (no, not that one, this one) and pretty soon you're back at the workbench, happily wailing away on your next *magnum opus*.

☞ ☞ ☞ HEALTH WARNING ☜ ☜ ☜

Our precursors gradually emerged from the primordial ooze to progress through the stone, bronze and iron ages. Since Aëtius defeated Atilla on the Mauriac Plain, our kin have brought about the Dark Ages, the Middle Ages, the Renaissance, the Age of Discovery, the Reformation, the Age of Enlightment and the Industrial Revolution. In our century, we've witnessed the dizzying passage through the Horseless Carriage, Electrical, Air, Jazz, Atomic, Electronic, Aquarius, Space and New Age. Having borne all of that with relative aplomb, we're now firmly, enmeshed in the Age of Litigation. Let's just hope we survive it.

One of its manifestations is an institutional preoccupation with Health and Safety, which has led to Sturgeon General's warnings on everything from beer to baggies. Not that anyone really cares more about our welfare than before – just about protection from lawsuits. *Caveat venditor.*

Nevertheless, there are certain real potential health hazards even to our hobby. Topping the list are certain highly volatile hydrocarbons which we routinely use for assembly and painting. And heading the list of toxic volatility is methyl ethyl ketone (MEK)*; a favored modeller solvent for polystyrene plastic. If inhaled extensively, it may cause damage to lungs, kidneys and the central nervous system. It's also highly inflammable. Fortunately, we usually decant the stuff from purchased cans into tiny glass jars, which are opened just long enough for application.

Perhaps of more actual risk is inhalation of paint/thinner fumes-lacquer in particular-during long airbrushing sessions. For this activity, wearing a good respirator is a minor inconvenience when compared to the potential risks.

For a number of years, I (Alcorn) often suffered from a vague yet distressing feeling which, for lack of a better term, I called "malaise." Otherwise, I was in excellent health, largely by virtue of running five miles, four times per week (for 34 years). Since doctors could find nothing wrong, I'm sure that they wrote me off as just another type A hypochondriac. I knew better, though I didn't know the cause.

In retrospect, it was most likely a steady dose of lacquer thinner fumes; from modelling, and later, auto restoration – I did all of my own painting. Belatedly, I began wearing a proper respirator, and eventually extricated myself from old car fever. As the bank account recovered and the wife's disposition improved, so did the malaise gradually subside. It had been no joke; *Deo volente*, no more sinister long-term effects of lacquer thinner abuse will appear.

Meanwhile, I'm now a true believer in respirators and proper ventilation for modelling-and eventual restoration of Gretchen the Thunderbug: my all-original 1956 VW "Kafer."

* Incidentally, for almost all references to MEK cited herein, commercial liquid plastic cement is an acceptable, if not preferable, alternative.

16 **SCRATCH BUILT!**

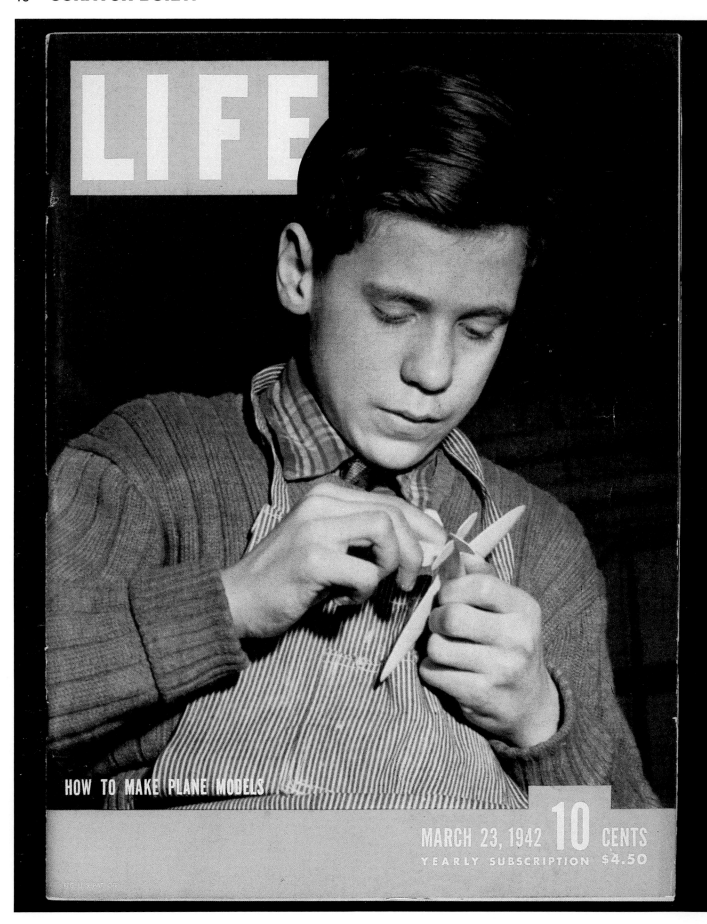

CHAPTER I: ROOTS
THE ORIGINS OF SCRATCH BUILDING STATIC SCALE MODEL AIRCRAFT

The roots of modern scratch building in polystyrene plastic go back to the era of "solid models", when kids of all ages fashioned the object of their fancy from wood.

Solid modelling became an industry in World War II with the federally sponsored program to mass produce 1/72nd scale recognition models in high school shops. Postwar, scratchbuilt solids soon gave way to preformed plastic. While this appeared to doom the requirement for hobbyist skill, it proved only the beginning — for kit bashers as well as scratchbuilders.

THE ANTEBELLUM ERA

While occasional examples could be cited from the earliest days of the airplane itself, the art as we "old-timers" know it began to flourish in the mid-1930s, with the advent of "scale" 3-view drawings, appearing in such journals as Flying Aces, Model Airplane News, and Air Trails. These pandered to the aviation interest kindled by barnstorming, Lindbergh, the growth of private and commercial aviation, nostalgia over the aerial heroics of the Great War, and the National Air Races.

In Britain, the doyen of solid scale builders was James Hay Stevens. His book, Scale Model Aircraft, published in 1933, described the rudiments of the craft and featured simple 3-views of a dozen aircraft, mostly WWI types. All were to 1/72nd scale, a recent choice by Stevens for Skybird kits (see below). Previously, he had used 1/36th scale, chosen for compatibility with his army of lead soldiers (based upon his assumption that the 2" high figures were meant to be six footers).

In a sense, solid scale modelling was simply the timorous stepchild of flying modelling, which by this time enjoyed a large, enthusiastic following, culminating annually in such epic events as the Wakefield Trophy and American Model Academy Nationals — and incidentally, providing youthful inspiration for the generation which was soon designing, building, maintaining, and flying the now legendary machines of the next War To End Wars.

During this same period, solid model kits appeared: basically a decorated box with 3-view "plans" and appropriately sized blocks of balsa, plus perhaps a cast lead propeller, rubber wheels and insignia — printed on slick paper for gluing

James Hay Stevens: The founder of our hobby. (via Harry Woodman)

to the finished surface. Really, each model fashioned from such a kit was "scratch-built." To our knowledge, the earliest truly pre-fabbed American solid models were those produced by Strombecker, who pre-formed the basic elements (fuselage, wings, nacelles, tail, etc.) from hardwood by some pre-"numerically-controlled" machining process. Among those fondly recalled classics from the late 1930s and early 1940s were their Bell Airacuda, Boeing Stratoliner and China Clipper.

Opposite: This evocative shot of a high schooler sanding on a 1/72nd scale Spitfire introduced an article within, covering the recently initiated recognition model program. We wonder: who is he; is he still modelling? (Reproduced courtesy Charles Steinheimer, Life Magazine © 1942, Time Warner, Inc.)

SCRATCH BUILT!

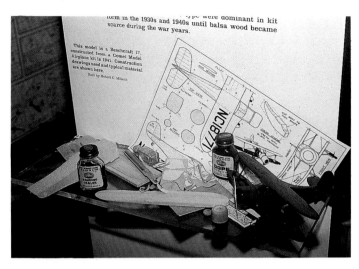

The Beechcraft 17 from this classic prewar solid kit is in fact "scratch built," since the balsa components were only profile cut - except for the lathe turned cowl and wheels. We older modellers can smell that Testors dope in our mind's nose. (Model and photo by Robert Mikesh)

Shown in this NASM display are two classic prewar Strom-Becker kits: the Sikorsky S43 and Boeing 314. (Mikesh photo)

Meanwhile, British modellers of the depression era who could muster two shillings (about 25 cents) were well served by the Skybirds "Givjoy" series, which began in 1932. These kits, of WWI and sundry contemporary types, all to 1/72" scale, featured general arrangement drawings, roughly pre-formed wood major components, cast lead propellers and engines, turned brass wheels, stamped metal struts and undercarriage legs, celluloid tail surfaces, gummed back paper markings and powdered glue – all packaged in lurid crimson-lined boxes featuring a lively James Hay Stevens drawing on the cover. Stevens, in fact, was the firm's "technical director", selecting the subjects, planning the kits, drawing the 3-views and cover art, and composing most of the textual material for the kits and bulletin.

The founder of Skybirds, toy merchant Alfred J. Holladay, was endowed with a distinct entrepreneurial flair. In 1933, he began publication of *The Skybird* quarterly bulletin (later to merge with Aero-Modeller) and founded the Skybird League, whose chapters about the country were organized through "District Commodores." All of this was very "clubby", with great emotional appeal to those of the swanscombe/beaker/celtic/roman/anglo-saxon/skylding-geat/norman blood line. Eventually, some 400 chapters sprung up throughout the Empire, from Canada to South Africa: their activities were chronicled in *Aero-Modeller*'s "The Skybird League" feature. Sadly, *Skybirds* folded its wings in 1945, having been severely restricted during the war years by materials shortages.

In 1933, *Skybirds* faced competition from the Frog "Penguin" series. Frog, in fact, was an acronym meaning "Flies Right Off the Ground", in reference to their rubber-powered flying model line. "Penguin", of course, referred to their flightless stablemates, which were moulded in acetate, tinted as

Box art, aircraft configuration and model construction were prepared by James Hay Stevens. (Model and photo courtesy Barry Gray via Harry Woodman)

This classic 30s kit included preformed wood major components, formed wire struts, printed insignia and powder glue - all mounted in a lurid crimson backing. (Model and photo courtesy Barry Gray via Harry Woodman)

deemed appropriate to the subject. Being first out of the chute with plastic, Frog emerged post-war as a pioneer of this genre.

WARTIME DEVELOPMENT

Aside from the War itself, surely the biggest single factor in the growth of solid scale aircraft modelling was the nationwide program of 1/72nd scale recognition model construction, initiated in early 1942 under the sponsorship of the U.S. Navy Bureau of Aeronautics. Almost overnight, kids all over the country – mostly high schoolers in shop classes – responded to the clarion call by whacking out Me109s, P-40Es, Spitfires, He 111s, Zeros, and such by the hundreds of thousands, in a veritable frenzy of patriotic wood butchery. The program included publicity, guidance for construction of the models, provision of plans for the various aircraft, and logistics for collection, selection and disposition of the models. Soon the plans, printed on heavy tan paper (suitable for cutting out as templates) were available commercially in boxed kits produced by Comet, which included hardwood blocks for the main components.

Since the models' role was in-flight recognition, only the basic shape was to be provided, augmented by major features of significant identification value, such as turrets, exhausts, intake ducts, spinners, and fixed landing gear when appropriate. They were to be completed in black paint, with no markings whatsoever.

In 1943, the program was supplemented by factory mass-produced models, molded in black cellulose acetate plastic (which by now are pricey collectors items, if not heat warped

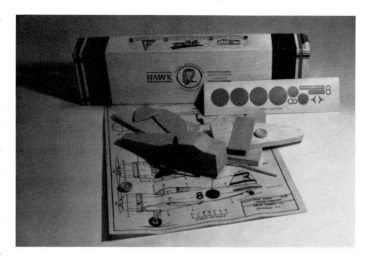

This classic early wartime solid kit features major components profile cut from balsa, lead generic propeller and paper insignia. The 1/48th scale "Zero" in fact represents the mythical Sento Ki-001 (see accompanying drawing). (Alcorn kit and photo)

into some grotesque caricature. Incidentally, after the War, the molds were acquired by Aristo-Craft of Newark, New Jersey, who marketed them for awhile). Doubtless this greatly reduced the logistics hassle of model gathering and resulted in items of uniformly high quality. Though surely cost effective, it deprived us kids of this means for direct participation in the War effort, aside from those old enough to attend the real thing. But meanwhile, the static scale modelling urge had reached epidemic proportions, which in various manifestations has raged ever since.

As the War progressed, the plethora of monthly journals pandering to the aviation/modelling enthusiast ran more and more scale plans features. Draftsmen such as William Wylam in Model Airplane News and Thomas A. Naylor in Air Trails produced detailed efforts on selected subjects. Despite the profusion of detail, the accuracy varied widely, apparently a function of source material, conscientiousness of the draftsmen and schedule constraints.

In 1944, Air Trails began a series of 1/72nd plans features which, though not greatly detailed, were of consistently good quality – reflecting customer sophistication as well as source material availability. Another important plans source for us scale enthusiasts during the later wartime period was the excellent Plan Pack 1/4 scale series drawn by Thomas A Naylor for Modern Hobbycraft.

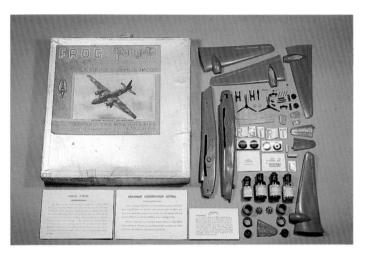

This late prewar/early wartime Frog "Penguin" Wellington serves to remind us that plastic kits as we know them have been around for over half a century! Granted that the surface detail is minimal and rather coarse, and the main components are molded in acetate rather than polystyrene: nevertheless, the overall configuration appears to be reasonably accurate, warpage notwithstanding; and wet transfer decals are in evidence, along with paint! US plastic kit manufacturers were 20 years behind this pioneering firm. Frog was originally an acronym for "Flies Right Off (the) Ground"; "Penguin" of course referred to the firm's earthbound products. (Cooke photo: kit courtesy Tony Bamford)

By this time, solid model kits had come a long way too, as epitomized by Maircraft's 1/4 scale line, featuring quality drawings, silhouette cut wood blocks and genuine wet transfer decals. In fact, Comet's 1/24th ("half inch") scale Speedomatic offerings, though ostensibly rubber powered flying models, served American youth primarily as "hanger queens", faithfully representing such popular stalwarts as the P-40C, Zero and Wildcat.

By 1945, several manufacturers were competing for the "youth market" with solids whose main components were preshaped, following the trend set prewar by Strombecker. These included Cadet, Testors ("No Carving! No Guesswork!"), Consolidated ("Redi-Carved") and Comet's 3/16"

SCRATCH BUILT!
PROCEDURE CHART FOR BUILDING SCALE MODEL AIRCRAFT

Chapter I: Roots

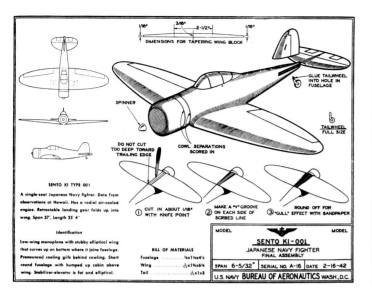

The mythical Sento Ki-001 appears to be the strange fruit of a union between a Zero and a Val. In fact, "sento-ki" simply means "combat aircraft." (Reproduced courtesy X-Acto ®)

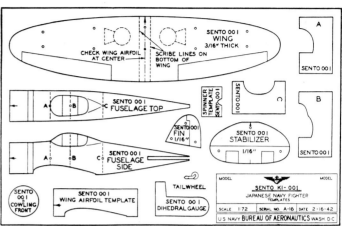

While the configuration of the Sento Ki-001 is ludicrous, the template sheet, printed on stiff tan paper, is typical of those for the recognition model series. (Reproduced courtesy X-Acto ®)

scale line, which ultimately featured a 26-1/2" wingspan Superfortress at $2.50 (that's like $25.00 today, folks).

THE RIGHTEOUS SOLID MODEL

Aircraft modelled in solid wood suffer from one great disadvantage: they can't be entered in IPMS contests. Despite this handicap, a properly rendered solid model can be every bit as worthy and impressive as its plastic counterpart. This point is well illustrated by a trip to the National Air and Space Museum, where numerous examples attest to the wood modeller's skill. Among the finest are those lovingly rendered by Syl Kill, including his wonderful Supermarine S6B to 1/16th scale (see photo).

Aside from being debarred from IPMS competition, solids suffer from other inherent disadvantages. Obviously, interior areas can only be modelled by hollowing the wood in that region, or by cutting away the entire section and inserting surface panels. Also, it is difficult, or worse, to scribe surface detail such as panel lines into even the smoothest of wood. Finally, even the finest, well cured wood tends to shrink somewhat over the years, resulting in paint cracks at component joints. But they are sturdy, dimensionally stable overall, and satisfyingly heavy when picked up.

The basic principles of producing the main components of a solid model are covered in Chapter III: Vacuforming, under the section Carving The Forms.

Desk Top Models: An area which tends to be overlooked, or perhaps looked down upon by serious modellers, is that of "desk top" display models. Not extensively detailed and mounted in a flight attitude upon a pedestal, they can be satisfying decor in places such as a living room or office. For such models, interior detail, retractable landing gear, and propellers are not required, and wood is perhaps the most expedient material. Consider the potential:

Your guests have arrived, been served their first libation and conversation begins to liven: opening gambits include the usual incisive observations upon weather, yesterday's ball game and the latest palimony suit. You have struck a noble pose, Jack Daniels in hand, elbow on the mantle, beside your latest creation.

The boss's new wife, Bonnie, comes up and gaily inquires: "Hi Ted, nice party, – oh! what a pretty airplane! Did you fly one of those?" Oh, god, you think, do I really look that old? But, quickly recovering your aplomb, you volunteer, "Uh, no. This was a Bell Airacuda, conceived in the mid 1930s to our erroneous strategic defense concept which envisioned waves of unescorted long range enemy bombers approaching our shores. See these manually operated 37mm cannon in the wing pods – that's a pusher Allison engine in back." Trying not to appear dumb, she speculates: "You mean somebody was in there! How would the poor boy get out past the propeller if something went wrong?" An awkward silence ensues while you search in vain for an answer. Endeavoring to recover the initiative, you point to the funny thing on the side of the pod: "The plane had many other innovative features, including these turbosuperchargers – in fact, the thirteen YFM-1s featured a tricycle landing gear . . ."

You look around, awaiting her response, hoping it won't be as dumb as her last one. But by now, of course, she's chatting airily with Dennis. He's a turkey – destroys perfectly good vintage cars to make street rods. But he's only 35 and has that shock of curly black hair.

Opposite: This "how to" chart for constructing recognition models was widely circulated in booklet illustration and poster form. (Reproduction courtesy X-Acto ®)

22 SCRATCH BUILT!

SUPERMARINE S6B by Syl Kill: This fine example of the solid modeller's craft now resides in the NASM. Syl was strictly of the old school, though he developed traditional methods to their limit. (Syl Kill photo)

Later in the evening, after having discreetly drifted off to do some fast research, you again find yourself next to Bonnie: "Actually", you say, "He could jettison the airscrew and jump through a hatch." She thinks: "This guy's really weird! I should speak to Kenneth about him." Maybe next time you should leave the model down in the den.

THE PLASTIC ERA

POSTBELLUM HIATUS: Really, it's only natural that solid modelling should have prospered, while flying modelling languished during the War. Aside from the impetus of the recognition model program, we youth of that era related strongly to the aerial engines of war, avidly following every operational and technological development as it unfolded (or was revealed) in daily news bulletins, weekly news magazines (*Life*,

TRAVELAIR "MYSTERY SHIP": Another of Syl's solid masterpieces now displayed in the NASM. (Syl Kill photo)

Time, etc.), the monthly aviation journals, and in the very skies above our heads. So, to replicate the objects of our fascination with a minimum of strategic materials was atune with the prevailing climate: gasoline, engine manufacturers, rubber, balsa (in quantity), and the guys who understood RC had gone off to war.

But thereafter, static scale modelling drifted into the horse latitudes as free-flight, control line and radio control prospered – spurred by product availability, new technology and hordes of demobilized enthusiasts. For example, the proliferation of gasoline engines in the USA was astonishing: Cannon, Delong, Super-Cyclone, Thor, G.H.Q., Torpedo, O.K., Fox, Hornet, Rogers, McCoy, Forster, Pac Meteor, Vivell, Champion, Bullet, Phantom, Atwood, Bantam, Herkimer, Atom, and others appeared (or, reappeared) to challenge the supremacy of Ohlsson and Rice.

SCRATCH BUILT!

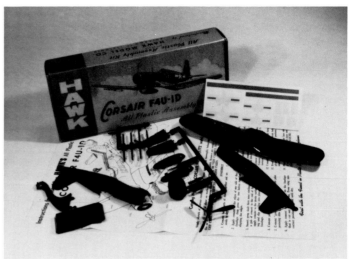

Here we see one of Monogram's transitional wood/plastic Superkits, introduced in 1952, during the Korean War. (George Lee photo)

This mid-1950s 1/72nd scale all-plastic kit is reasonably accurate and crisp in overall configuration, though dead simple - the wings are solid! The box art and exploded assembly drawing are dreadful! (Alcorn photo)

Meanwhile, Hawk, Maircraft, C-Z and a few others continued to offer solids to a dwindling market. Many of the serious, war smitten cases, including George and I (Alcorn) eschewed kits in any event, preferring to simply work from drawings, which continued to emanate from the pages of the modelling journals. We had long since come to regard balsa as mushy kid stuff and were downright contemptuous of pre-formed parts. It was during the late 40s and 50s that George produced an extensive series of 1/72nd scale World War I models, which even today are gratifying to see.

PLASTICUS VINCIT OMNIA: Then, like a new strain of bacillus, a foreign substance began to invade certain kits – plastic! In 1952, Monogram – a Johnnie-come-lately among manufacturers-introduced a line of 89 cent balsa "Superkits" sporting plastic parts, including propellers, radial engine/cowlings, canopy, armament, display stand – and bust of the pilot! Initially, three choices were available: Corsair, Mustang and Thunderjet. As their ad proclaimed: ". . . Even with all the detail, building is so easy, you can't do it wrong. Sm-o-o-o-th finished balsa, plastic parts and genuine authentic decals."

Meanwhile, Allyn carried the concept to its ultimate extreme, with a series of (mostly) jets entirely of plastic. While reasonably accurate in overall configuration, and pleasing as desktop decor, they were the acme of simplicity – literally, they could be assembled in minutes: even their correct single color was cast into the material. Their subjects, pedestal mounted in flying attitude, were selected for minimal complexity: the Douglas Skyshark, Skystreak, Skyrocket and Skyknight, plus Boeing Stratojet. Hawk followed suit with simple pedestal mounted, gear retracted offerings of standard subjects. By 1955, Monogram had made the transition to all-plastic kits, introducing a DC3 in TWA livery, replete with scale pilot, copilot, hostess and passenger. This line of "Four Star Plastikits" included a B-25G, B-26 (A-26) and "Old Dumbo," a PBY-5A-all with landing gear extended. Meanwhile, as related earlier, the Brits had long since converted to plastic, with extensive offerings from Frog, Airfix and Merit.

Surely the era of static scale modelling – as opposed to assembly – was ending. It seemed that the attentions of serious practitioners had turned almost entirely to flying models, many of which were "exact scale." Indeed, the March 1958 *Model Airplane News* contains not a single ad for static scale models – either wood or plastic. (Incidentally, this issue is something of a collector's item, since it features scale drawings of the "super secret" Lockheed U-2 from Clarence Kelly's now legendary Skunk Works. It would be two years before Francis Gary Powers' misfortune would stun the world by the widespread revelation of its existence and of our "covert" surveillance of the Soviet Union. Bjorn Karlstrom's reasonably accurate, two-year old drawings added to the discomfort of our clandestine activities establishment.)

Then, almost imperceptibly, the worm began to turn. In December 1960, MAN ran a tiny ad for 1/72nd scale Airfix kits: by February 1961, it was almost full page. Monogram was still there and before long Asian Sprue (as opposed to Asian Flu, which I helped introduce in 1956) had reached our shores. You know the rest.

PLASTICUS COGNITUS: So, just what is this stuff we call plastic? (20 questions) Is it organic? Yes, in the sense that it's made primarily of carbon, hydrogen, oxygen and nitrogen. Is it natural? No, synthetic (uh, oh). Is it biodegradable? Yes, in acetone, toluene and MEK. Is it safe? It shouldn't be taken internally. Is it sexist? Uh, no-it can be formed in either male or female molds. Is it racist? Uhh-er – it's mostly white. Is it ethnocentric? Ummm. While most of the end product subject matter is culturally northern European, most of the profit goes to Asia. Is its use addictive? Yes, although epoxy can be cured. Does its use constitute substance abuse? We've seen some fairly serious examples, even at national contests (Just say "no" to visible seams and fantasy subjects). Let's stop here.

Plastics can be broadly defined as any long chain (polymeric) hydrocarbon compound which in the heated state can be molded, extruded, cast, rolled or drawn; but which can be made to solidify at room temperature. In a general sense, plastics are usually categorized as follows:

A. Cellulose derivatives, including cellulose nitrate (celluloid) and cellulose acetate ("dope");

B. Other natural plastics, including casein (from cows); shellac (from bugs); amber, rubber and gutta percha (from trees);

C. Synthetic resins: By convention, the industrial definition of plastics is confined to this broad class. These substances can be categorized either as:

• Thermosetting: Upon heating and/or catalyzing, the molecules crosslink (cure) into a 3D network, which cannot be heat reversed, or;

• Thermoplastic: The long linear molecular chains do not crosslink, so that the material can be reversibly reheated following initial solidification (polymerization). They form an amorphous mass of entangled long chain molecules: picture a can of really sticky worms.

These resins can in turn be classified by the chemical means employed to achieve the polymerizing reaction, namely;

• Addition polymerization; in which the long chain molecules (polymers) are formed from their monomer constituent without change of chemical composition (no release of volatiles) or;

• Condensation polymerization; wherein monomers react to link, accompanied by release of volatiles, namely water and alcohols.

Combining these categorizations, we have:

• Addition polymerized thermosets, such as epoxy;

• Condensation polymerized thermosets, such as Bakelite (phenol-formaldehyde);

• Addition polymerized thermoplastics, including acrylics (Plexiglass, etc.), polystyrene and PVC (polyvinyl chloride);

• Condensation polymerized thermoplastics, including Celluloid.

While we tend to think of plastics as modern-along with fast electronics, nuclear power, space travel and environmental impact reports-their laboratory synthesis goes back a surprisingly long way.

Nitrocellulose was synthesized by a French chemist in 1833. The molding of plastics; that is, the introduction of a pourable resin into a receptacle of the desired final shape, within which it could be solidified, was first achieved by Charles Goodyear in 1839, with his vulcanizing process for rubber. But, that's still a natural substance. The first commercially important synthetic plastic was cellulose nitrate – Celluloid, developed in 1869 by John Hyatt. Though highly flammable, it soon found widespread application as a substitute for expensive ivory in production of billiard balls, piano keys, false teeth, "tortoiseshell" combs, buttons and such. Sheet celluloid

Chapter I: Roots

A PLASTICS PRIMER

ADDITION POLYMERIZATION: Identical monomers link without change in chemical composition.

Example: ETHYLENE: $\begin{matrix} H & H \\ | & | \\ C = C \\ | & | \\ H & H \end{matrix}$, a monomer with a double carbon bond

can link to form the polymer

POLYETHYLENE: $\left[\begin{matrix} H & H \\ | & | \\ -C - C - \\ | & | \\ H & H \end{matrix} \right]_n$

This reaction may be written
$CH_2 = CH_2 \rightarrow [-CH_2 - CH_2-]_n$, where $n \geq 1000$

Polymerization is facilitated by a catalyst, resulting in an exothermic reaction with no release of constituent volatiles.

CONDENSATION POLYMERIZATION: Bifunctional monomers react to form a chain, releasing water and/or acids. Such "bifunctionality" operates through reactive groups such as:

HYDROXYL ($-OH$); ACIDIC ($-COOH$); or AMINE ($-NH_2$).

EXAMPLE:

$H-N-C-C-C-C-C-C-N-H + H-O-C-C-C-C-C-C-O-H \rightarrow$ (releasing H_2O)

hexamethylene diamine   adipic acid

$\left[-N-C-C-C-C-C-C-N-C-C-C-C-C-C- \right]_n$

NYLON, a polyamide

(Susan Esp, per Alcorn)

THERMOPLASTICS: Following post-polymerization solidification, thermoplastics can be replasticized by heating, without chemical change. All linear polymers are thermoplastic.

Example: POLYSTYRENE: $\left[-CH_2 - \underset{\bigcirc}{CH} - CH_2 - \underset{\bigcirc}{CH} - CH_2 - \underset{\bigcirc}{CH} - \right]_n$

where $\bigcirc$ = benzene ring (C$_6$H$_6$)

Example: POLYVINYL CHLORIDE: $\left[-CH_2 - \underset{Cl}{CH} - \right]_n$

THERMOSETS: Polymerization proceeds in chain and cross links to form a 3D network. After polymerization, thermosets cannot be replasticized without suffering chemical change (degradation).

Example: BAKELITE: phenol + formaldehyde $\rightarrow$ ...

The three hydroxyl ($-OH$) groups can react with similar monomers (releasing H_2O) to form a condensation polymer.

SCRATCH BUILT!

This is a selection of 1/72nd scale W.W.I subjects made by George Lee during the 1950s. (Ben Walker)

provided shirt collars, side curtains for buggys (later automobiles) and motion picture film. Bakelite, developed in 1910 by Hendrik Baekeland (a Belgian-born U.S. chemist) found wide application as an electrical insulator and material for telephone (and other) housings.

We now present some amazing facts about our plastics:

- PVC: Vinyl chloride, a gas at room temperature formed by the reaction of acetylene with hydrochloric acid, was first made in 1835. The polymer, polyvinyl chloride, was first produced in 1912;

- Acrylics: plexiglass-polymethyl methacrylate resin – saw widespread use in World War II as cockpit, turret and nose transparencies, as we all know;

- Polystyrene: Styrene is vinylbenzene, first synthesized in 1839. Chemically, it is similar to natural rubber. Our material – high impact polystyrene, is toughened with butadiene rubber.

IPMS

We all owe the IPMS a great debt for providing the overall context within which we pursue our hobby. In fact, it's difficult to recollect modelling before IPMS. As related above, we had a few magazines which pandered to the solid (and early plastic) static scale aircraft modeller: otherwise, we were on our own except for like-minded friends and chance encounters with kindred spirits (such as mine with George Lee, at which time I was only dimly aware of IPMS). Model contests? Oh, sure, local hobby shops would occasionally host such events as a form of promotion. I (Alcorn) recall winning a Rogers 29 gasoline engine with my kit-built Grumman F7F in 1946: still have the Tigercat, but I was soon hornswoggled out of the engine by an acquisitive schoolmate, who had the wit to perceive that I had little use for the engine – and no idea how to make it run.

But, now we have the support of an entire industry, plus our very own society, with its mystic rites; bylaws; journal;

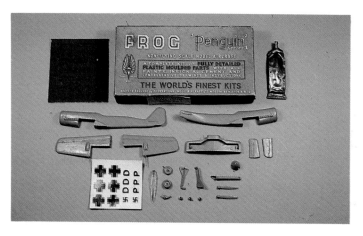

Listed simply as an Fw 190 among their early postwar offerings, this is clearly a Jumo powered Dora. Overall quality is comparable to their prewar line. Frog failed to prosper postwar, being eclipsed by Airfix which featured injection molded polystyrene components. (Cooke photo: kit courtesy Tony Bamford)

chapter, regional and national conclaves – the latter rivalling those of other professional organizations in terms of attendance, enthusiasm, seminars, vendor displays and awards banquet ritual.

Like so much of our cultural heritage, the origins of IPMS go back to the Mother Country. As explained by Jim Sage in the September 1991 IPMS/USA Journal, it was a visionary named Peter Elley who in 1963 conceived the idea of a British Plastic Modeller's Society. Shortly after his brief but seminal involvement, the embryonic organization had relinquished its insular title for International PMS. Having violated the sacred military dictum of "never volunteer for nuthin'," Sage founded the USA branch of BPMS, which inadvertently forced the change of title to IPMS.

He goes on to relate how the first "convention" in October 1964 was really no more than an informal gathering of about 25 Chicago area modellers, who met one Sunday in a Des Plaines restaurant. Miami in 1967 was the first multi-day hotel affair: by 1970 the IPMS/USA National Convention had evolved into the extravaganza which we know and love so well.

BELLANCA "CRUISAIR" FLOATPLANE: Canadian Ron Lowry's impressive Bellanca took first in its scratchbuilt category at the 1987 IPMS/USA Nationals in Washington, DC. It now graces the National Aviation Museum in Ottawa. (Lowry photos)

CHAPTER II: HARD CORE LORE

COMMERCIAL PLANS SOURCES

MAP: Ever since the War, a primary source of quality multi-view drawings has been the British journal *Aeromodeller* and, since 1970, its offspring *Scale Models*, produced by Model and Allied Publications (MAP) of Hemel Hempstead, Herts. For decades, Aeromodeller ably served both the flying and static scale enthusiast, under the leadership of Ron Moulton. Through the 1970s, *Scale Models* was blessed by the inspired guidance of Ray Rimell, a first class model builder/aviation historian in his own right. From the 1950s into the 1980s, their staff of fine draftsmen built up the impressive catalogue of plans available today through Argus Specialist Publications, which acquired MAP in 1984. These dedicated individuals included Dave Platt, Ian Stair, G.R. Duval, Pat Lloyd, Ken Merrick, Doug Carrick, R.H. Cooksey, G.A.G. Cox and Arthur Bentley. Bentley produced the superb series exemplified by Plan Pack 2993, delineating the Focke Wulf Fw 190 A, F, G in five sheets of exquisite detail. The Hawker Hurricane, Typhoon, and Tempest received this same treatment.

Incidentally, I (Alcorn) had the satisfaction of providing background drawings for Bentley's multisheet feature on the long neglected Douglas DB-7/A-20/Boston series. An equally satisfying sidelight is that my friend, the redoubtable Group Captain James Pelly-Fry, constructed a 1/8th scale, radio-

Here we have Bob Rice gingerly holding his epic Il'ya Muromets. (Arizona Daily Star photo)

Chapter II: Hard Core Lore

controlled Boston III from these plans. His model represented the aircraft in which he, as 88 Squadron C.O., led the famous low-level Boston/Ventura mission against the Philips Valve Works at Eindhoven in December 1942 (60 A/C out, 45 back).

Argus Publications has recently (1988) initiated a series of topical monographs, *Aircraft Archives*, featuring reprints of selected MAP plans supplemented by photographs and concise text.

While the MAP/Argus output has provided the foundation of widely available material for the serious scale modeller, numerous other commercially published sources have made valuable contributions.

MAN: For a long time after the War, *Model Airplane News* continued to provide plans for the scale modeller, which included profusely annotated epics by William Wylam. Arguably the finest drawings ever to appear in MAN were those produced by Bergen Hardesty on the Nieuport 17/24/27/28 series.

Paul R. Matt: Our debt to Paul is acknowledged by the dedication of this book. His HAA drawings have been the basis for many a scratch built project. (Photo courtesy Sun Shine House, Inc.)

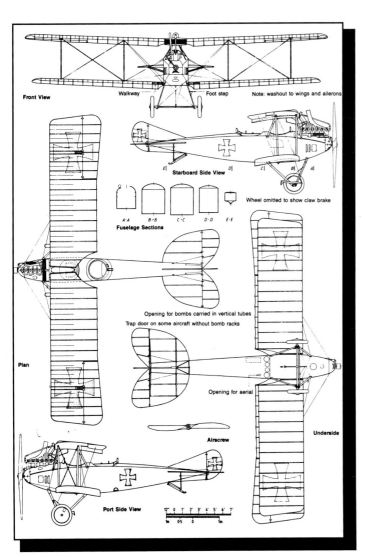

ALBATROS C.III: These Ian Stair produced drawings of the doughty WWI German two-seater appeared in Windsock Datafile No. 13. They are representative of the many fine efforts by Ray Rimell's Albatros Productions. (Courtesy of Albatros Productions)

AIR PROGRESS: Postwar editions of this interesting U.S. publication had something for everyone of the aviation persuasion, including pithy features on vintage aircraft by Peter M. Bowers and (during the 1950s) copious scale views by Walter M. Jeffries, who later achieved fame and fortune in the movie industry. His efforts were invariably shaded to achieve a realistic 3-D effect, unlike those of his successor contributor of the 1960s, Bjorn Karlstrom.

SUPERSCALE: In the early sixties, E. R. Atkins, Jr. of Arlington, Texas marketed the now legendary *Superscale* series, supplied in large (1/16th scale) blueprint format. With one or two exceptions, they set new standards for accuracy and profusion of detail, both interior and external. Two of these efforts were landmarks which have yet to be excelled for the types represented: the Lockheed P-38L by Leroy Weber and the Supermarine Spitfire MKI by G. A. G. Cox and M. J. Lee – rights to the latter were evidently later acquired by Aeromodeller, as the basis for their Plan Pack No. 2896. Other notable efforts in this all too-brief series included the North American P-51B Mustang, by Atkins himself, and the Mitsubishi A6M-5 Zero, by Kukuo Hashimoto. Needless to say, both the large format blueprints and a soft cover book featuring all of the earlier Superscale productions in reduced size are now jealously guarded collector's items. However, reprints of the Superscale drawings are now available from Sunshine House of Terra Haute, Indiana (see HAA below).

HISTORICAL AVIATION ALBUM: Of the many historical aviation enthusiasts who have contributed to the lore of the scale modeller, Paul Matt stands supreme. Within the frail breast of this quiet, gentle man burned a lifelong passion for preserving our aviation heritage. A fine "solid" modeller himself (and PBY aircrewman during the War), Paul devoted his career to creating a series of publications of enduring quality, which are among the most treasured documents of any devotee of American aviation history. Within each of the volumes of his *Historical Aviation Album* were meticulously

SCRATCH BUILT!

This model by Bob Rice represents one of the 90 examples of this parasol-wing observation type supplied to the Army Air Corps and National Guard Units during 1936/37. (Rice photo)

researched, well written and artfully presented features on an astounding variety of aviation subjects. His interest was as eclectic as his output was prolific. From the Timm Collegiate to the North American XB-70 Valkyrie, once a subject had been selected for scrutiny he was unrelenting in his search for documentation and perspectives. Paul's most personal and direct contributions for the scratch builder were of course the labors of love which emanated in great profusion from his drafting board. Even in his last years, when arthritis had painfully stiffened the swollen joints of his hands, Paul persevered: the drawings of the Taylor/Piper "Cubs", Curtiss B-2 and B-20 "Condor" and Curtiss F9C-2 "Sparrowhawk" which appeared in his last (and perhaps finest) HAA (Vol. XVIII) betrayed no evidence of his affliction, or of his failing health in general.

While HAA was essentially a one man (plus one woman, his wife Joan) publishing house, Paul drew upon many contributors for his articles: indeed, one of his great strengths lay in his vast web of aviation contacts, most of whom were loyal friends and admirers. In fact, his two principal contributors, Ken Rust and Thomas Foxworth, acted as Editor and Editorial Assistant, respectively, for HAA. Rust performed a prodigious feat of scholarship in preparing the series published by HAA documenting the operational histories for each of the overseas army air forces in WWII, from the 5th through the 21st.

I (Alcorn) enjoyed the privilege of providing the text, photos, data and background drawings for his two-part HAA series covering development, configuration and operations of the Douglas A-20/Boston. (Vols. XV and XVI).

Naturally, many an award-winning scratch-built aircraft model has been based upon Paul's drawings: witness George Lee's 1/16th scale Sikorsky S-39CS "Spirit of Africa" which now resides in the National Air and Space Museum (the superbly documented article and plans for this exotic leopard/bird appeared in HAA Volume XIV, along with the definitive tome on Benny Howard's "Mister Mulligan" and DGA series).

We and generations of aviation enthusiasts/modellers to come owe Paul Matt a great debt: in a very real sense, his spirit lives on in the pages of the *Historical Aviation Album* and his many other productions.

Paul folded his wings in November 1987; and as is so often the case with devoted couples, Joan followed him shortly thereafter. The assets and publishing rights of HAA were acquired by SunShine House, Box 2065, Terra Haute, Indiana, 47802. They have recently (1991) published a fine two volume set of soft-cover books: Paul Matts' Scale Airplane Drawings, covering 124 subjects from his HAA plans output, in 8-1/2 x 11 format.

WINDSOCK DATAFILE: Ray Rimell has brought manna to the masses of long-suffering WWI aircraft enthusiasts with his excellent *Windsock Datafile* series. Each 8-1/4" x 11-3/4" Datafile, prepared by a renowned WWI historian, provides prime photo coverage of a given aircraft type, complemented by a brief historical text; lively cover painting by Brian Knight; definitive multiview drawings to 1/72nd and 1/48th scale; and colour profiles by Ray himself.

JAPANESE CONTRIBUTIONS: Starting in the 1970s, the magazine *Koku-Fan* featured drawings by Kukuo

Chapter II: Hard Core Lore

Hashimoto and others, delineating most of the WWII Japanese types, as well as those of other nations. Airview responded with comparable efforts by Watanabe, creator of the awesome color profiles and supporting artwork which have provided the basis for Crown Books' huge format monographs.

But, from the modeller's standpoint, the ultimate Japanese contributions have burst forth in the pages of a journal with the decidedly inappropriate title of *Maru-Mechanic* – sounds like a trade journal for tramp steamer erks to us. Such worthy subjects as the Mitsubishi A5M and A6M, Kawasaki Ki-61 Hein, Nakajima B5N "Kate" and A1CH1 Type 99 "Val" have been presented to a level of authentic detail unlikely to ever be excelled. To round out the feast, each drawing has been included in an extensive presentation, featuring cockpit, engine, structure, landing gear and armament details, operational and detail photos, markings, illustrations and supporting text (unfortunately, only in the native tongue).

SQUADRON/SIGNAL: This worthy organization has initiated a series of monographs, and topical volume collections thereof, under the title Aerodata International. Development and operations of each subject are covered in a concise, well researched, written and illustrated feature accompanied by multi-view external drawings prepared by Alfred Granger.

Little mention is made of factory, "measured from life", service manual data or other documentation as the basis for drawing preparation (ex: since Brewster has long since ceased to exist and no examples survive, where did they find reliable dimensional/configuration data on the Buffalo?) While most of them published so far "look right", it would be comforting if some source acknowledgement were given for each. (It is analogous to presenting a scientific paper without reference to experimental data or relevant published material.) Nevertheless, they appear to represent a major contribution to the WWII modeller/enthusiast by providing quality drawings for several types previously inadequately covered, including the B-17G, TBD, F4F and F2A. We hope that this series will prosper, and that drawing prints to larger scales (1/4, 1/8, 1/32) will be made available.

OTHER RECENT SOURCES: If the Albatros DVa is the aircraft of your choice, then the Smithsonian's monograph, *Famous Aircraft of the National Air and Space Museum* #4, is

"SPIRIT OF AFRICA": George Lee's S39CS, c/n 914, NC52V – shown here in its NASM display case – represents the famous aircraft in which Martin and Osa Johnson made an epic 60,000 mile tour of Africa in 1933/34. Both George and Bill Bosworth (for his Varney Air Ferries subject, shown elsewhere) used Paul Matt's outstanding drawings of the S39B, which accompanied a long feature on the type in Historical Aviation Album (HAA), Volume XIV. (Mikesh photo)

SCRATCH BUILT!

Here Ray Rimell is preparing a color plate for the Windsock Datafile Special on the Fokker Dr.I. Ray is a cult hero to us W.W.I aircraft enthusiasts, for his conscientious dedication to authentic documentation of those primordial engines of war. Publisher, editor, historian, artist, modeller, Ray does it all. (Photo courtesy Ray Rimell)

your bible. Aside from curator Robert Mikesh's superb, indepth coverage of restoration of their example, they have reproduced Australian Bob Waugh's drawings based upon thorough measurement of the Canberra machine. It's all the information you'll need to build the real thing – assuming that you've already obtained a Mercedes D III engine from the local junkyard.

There must be other noble efforts available out there. For example, we haven't kept up with publications from France, Italy, Germany, Scandinavia or the Eastern European Nations.

FREE LANCE DELINEATORS: Before we depart from this discussion of commercial plans sources, we must mention an important but mystically obscure group of contributors who serve our cause by preparation of scrupulously authentic drawings of vintage aircraft; not primarily for financial remuneration but rather from a passionate interest in the machines themselves. To them, time and effort seem to be of no concern – only the psychic satisfaction of capturing the essence of some aircraft which has caught their fancy, yet which has not hitherto been rendered to their exacting standards of accuracy and detail. Eventually these academic tours de force may appear in commercial guise, but to them it is almost incidental.

Doug Carrick is an archetypal member of this Silent Service. Long time modellers/aviation enthusiasts may recall his contributions through MAP Plan Packs on the Messerschmitt Bf 109E (#2790), Bf109F (#2945) and Heinkel He111H (#2926). More recently, his extensive research of the Bf110C has borne fruit in Ken Bokelman's Scalecraft plans series (P.O. Box 4231, Whittier, California, 90607). Another recent Carrick contribution has been configuration material for the SPAD XIII in Rimell's Windsock Datafile on the type. His eclectic taste has also focused recently upon the Curtiss F8C-4 "Helldiver", Martinsyde "Buzzard", and Nieuport-Delage 29C.1. I (Alcorn) received invaluable support from Doug during lengthy preparation of DH9A fuselage structure/interior drawings, to supplement his fine exterior views as the basis for my forthcoming model of the type in 1920s "Empire" livery.

To Doug, Peter Westburg, Bergen Hardesty, Gene Schmidt, Joseph Nieto and others who have followed this noble calling, we owe a great debt. Incidentally, full size prints of Peter Westburg's 1/12th and 1/10th scale drawings can be obtained from Model Plan Services, 34249 Camino Capistrano, Capistrano Beach, CA, 92624.

A general admonition regarding all published plans: never take any as gospel without a thorough check against all available sources.

CONSOLIDATED NY-2: This NASM model of the pioneer blind flying research aircraft was made by Robert Mikesh. (Alcorn photo)

ARCHIVAL PLANS/ CONFIGURATION SOURCES

Chapter II: Hard Core Lore 33

FACTORY MATERIAL: In a word (or phrase), it is tough to obtain archival material directly from the manufacturer of an aircraft produced 40 or 50 years ago. Face it, there are precious few companies producing aircraft during WWII who are still in the business today. For those long since defunct, we're lucky if at least some of their archives have been acquired by museums or other institutions: Curtiss is a happy example, much of whose archives have been acquired by the Smithsonian's National Air and Space Museum. What with the many takeovers in recent years, many honored old names have been unceremoniously absorbed into larger conglomerates. Witness Douglas into McDonnell-Douglas, or North American into Rockwell.

Typically, even those few survivors usually tend to discourage the amateur from snooping through their files, even if the material is still there. It's a bloody nuisance, isn't cost effective, may be perceived (or rationalized) as a security threat, and probably doesn't interest them (most automobile companies are just as bad). Worse yet, older material of no possible utility for present business has often simply been pitched, in the interests of space demands, building renovation, or desire to eliminate the associated clerical costs. One of the most blatant, appalling cases which comes to mind is that of the venerable Packard Motor Car Company. As the story goes, when James Nance took the helm of the financially troubled concern in 1952, his "house cleaning" included destruction of all its historical archives. This doubtless also included corporate files on the WWI Liberty and WWII Merlin engines.

As "Götterdämmerung" approached in 1945, all of the German aircraft manufacturers were directed to destroy their records, both past and current. From this distant view, it's difficult to imagine the motives for that dictum – but then, after all, few can plumb the twisted rationale of the leaders of the Third Reich.

Of course, there have been some heartening exceptions: for example, we understand that (surviving) British firms are less disinclined to cooperate with the serious enthusiast than their U.S. counterparts. In our country, much valuable material

ZEPPELIN STAAKEN R.VI: This fine model of the awesome Riesen Flugzeug was constructed by Ray Rimell from the Contrail vacuform kit. (Ray Rimell photo)

has been saved, or at least retrieved by enthusiasts within the company. Often, these have been long-term employees, well aware of the historical importance of their company's products. For example, a dedicated faction at Douglas, including Harry Gann, have managed to make material available for serious endeavors, both private and commercial. Dustin Carter, also an AAHS stalwart, rendered similar service for North American, while Peter M. Bowers has made a lifetime crusade of perpetuating the many fine products of Boeing.

One problem with old drawings is often one dimensional shrinkage of the linen or vellum upon which they were drawn. I (Alcorn) experienced such difficulty with a (retraced) side profile of the DH9A, while Peter Cooke encountered this effect on Spitfire drawings at Hendon.

SERVICE MANUALS: Ever since the 1920s, a contractual requirement for production military aircraft supplied to the U.S. Armed Forces has been provision of an appropriate service manual, later accompanied by flight instructions. These Erection and Maintenance Manuals, as they are officially titled, are often of immense value to the scale modeller, although the quality of content from his perspective can vary widely among manufacturers or even specific products.

Let's take the example of T.O. No. 01-40-AE-2, Erection and Maintenance Instructions for the Douglas SBD-3, dated 3 January 1942 (Revised 30 July 1943) – George Lee just happens to have an original copy. Of primary value to the delineator or modeller is the extensively dimensioned multi-view general arrangement drawing. While such factory drawings are often suspect (or worse) as regards contours, this one is quite good. In any case, the dimensions permit verification of most aspects. Doubtless, this GA drawing (or the similar one for the SBD-5) was used as source material by Dave Platt for his excellent MAN Plan Pack SH 2872 on the type.

Also included is the canonical "Stations" drawing, giving locations of the various fuselage bulkheads and flying surface ribs. While this information is a useful adjunct to the dimensioned general arrangement drawing, the A/C contours are rather primitive. In fact, station drawings as a rule are suspect as to shape and proportions. Since they were prepared for a single purpose, little effort seems to have been expended in the interests of future modellers.

Second only in value to the dimensioned general arrangement drawing are the copious large-format photographs, showing cockpit, gunners compartment, landing gear, engine mounting, ailerons, dive brakes, arrestor hook, armament, bomb fork and such like in exquisite detail. Some 75 such photos in fact comprise the core of the document, presumably on the basis that "a picture is worth a thousand words."

Interestingly, the later E and MM on the SBD-5 is not quite as valuable to the modeller, containing few large format photos and a less well dimensioned 3-view. However, its many line drawings, often "exploded views", are of interest and were doubtless of considerable value to maintenance folk.

I (Alcorn) drew extensively upon E and MI's for preparation of my A-20 drawings. Chief among these were T.O. 01-40 AL and AD (A-20A), -40A (A-20B), -40AF (A-20C), and -40AL (A-20G, K). While I never managed to obtain an original, good copies came from a variety of sources.

I (George) used T.O. 01-25G dated Oct. 31, 1933 as the primary source document for my Keystone project, supple-

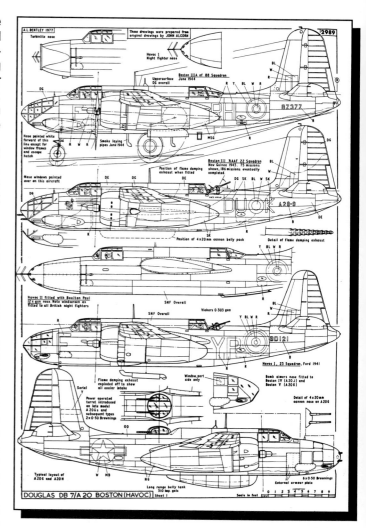

This is Sheet 1 of MAP Plan Pack No. 2989, which appeared in the 1978 Scale Models. I (Alcorn) had the honor of providing the background drawings for this feature. A.L. Bently traced and reformatted my drawings, adding the DB7/Havoc I, DB7A/Havoc II, Turbinlite and Boston IV/V details. (Copyright Argus Specialist Publications)

mented by many fine 8 x 10 glossy detail factory photos. Most were kindly supplied by Winthrup Shaw and Phillip Edwards of the NASM, Bob Cavanaugh of the Air Force Museum, and Bill Larkins. Paul Matt's Keystone feature in HAA, Vol. 5, No. 2, Summer 1972, was also useful.

Air Service Caravan of New Bedford, Massachusetts offers an extensive catalog of aircraft and engine technical manual copies, covering the modern to vintage products of most U.S. manufacturers and some British. Until recently, this otherwise splendid service was marred by the fact that they did not use a tone copier: hence photo reproductions were virtually useless. However, we are pleased to report that their products are now of acceptable quality for modeller reference.

VINTAGE "TRADE JOURNALS": Many a rich vein of hardcore lore resides between the yellowing pages of (usually) monthly publications which catered to the practitioner of the time rather than to the amateur aviation enthusiast. By now, such ephemeral journals are truly archival, being acces-

Chapter II: Hard Core Lore

DOUGLAS A-20A: This 1/32nd scale model represents a specific strafer-modified aircraft, "Daisy Mae", of the 89th Squadron, 3rd Attack Group, 5th Air Force, operational in eastern New Guinea (Papua) during the period September 1942 – late 1943. Though extremely effective in the ground attack role, the 89th's greatest claim to fame was its participation in the epic Bismarck Sea action of 3/4 March 1943, during which a 21-ship Japanese convoy bound from Rabaul to Lae was almost annihilated. The 3rd's other aircraft during this period were similarly modified B25 Cs and Ds. By 1944, the entire group had been equipped with the A20G. To the abject amazement of its creator, this model was voted Best of Show at the 1974 IPMS/USA Convention in Anaheim. (Joe Faust photo)

sible only through museums, libraries, aviation historians – or rarely through a chance junk shop/garage sale find.

Journals such as *The Aeroplane*, *Flight*, *L'Aerophile*, during the 1915-1919 era often featured in depth articles on specific World War I types which have come to be among the best sources for dimensional/configuration/performance information. Typically, the best features in the Allied journals covered certain captured German aircraft which had been subjected to thorough evaluation, both flying and static. For many important types of which no example or factory archives survive, this represents the best information available to the researcher. Such is probably the case of the Albatros CV/16 reconnaissance/bomber, covered in a 4-part Flight series from February 18, 1918 through April 11. While the GA 3-view of 28 February is useful only for a few overall dimensions, it was supplemented by excellent bulkhead cross-sections, first-rate dimensional structural drawings of fuselage, wings, and tail surfaces, O.A. photos and detail line drawings – to say nothing of the very thorough text.

A similar situation prevailed during WWII. In the U.S., the journals *Aviation* and *Air Tech* carried feature articles on selected axis and allied types, replete with photos, specifications, line drawings, and dimensioned 3-views. As discussed elsewhere, I (Alcorn) found the January 1944 *Aviation* feature on the Douglas A-20 to be an especially valuable source for detail/structural information. These in depth design analysis features in Aviation began with coverage of the Bell P-39 Aircobra in May 1943. Their March 1944 coverage of the Bristol Beaufighter carried an especially fine fuselage side view drawing showing full internal structural/equipment details. A similar trade journal, Aircraft Production, was published in England.

MUSEUM ARCHIVES: For the dedicated modeller/historian, the archives of major aviation museums around the world represent the ultimate document resource, although the personal files of some individuals can be tough competition in certain respects.

Certainly, museums are the logical repositories for vintage aviation journals, tech manuals, and books. As a rule, they have a great deal more. Photographs are an especially prized resource of certain museums: indeed, some have photo services of immense value. The Imperial War Museum in London, for example, has a magnificent aviation photo file from which high quality prints are available. Other valuable photo resources include the RAF Museum, the National Air & Space Museum, the Air Force Museum, National Archives, the Musee de L'Air, Establissement Cinematographique et Photographique des Armees (ECPA), Bundesarchiv, and British Ministry of Defence (MoD).

Museums can be unique repositories of certain regional or manufacturer-specific material and the beneficiaries of extensive personal collections having particular subject emphasis. A case in point is the "new" San Diego Aerospace Museum. There, you will find a rich trove of material on San

SCRATCH BUILT!

CONSOLIDATED P2Y-2: Arlo Schroeder's P2Y-2 was based upon Paul Matt's drawings. Flying boats make tempting subjects for scratch-builders due to their relatively poor kit coverage, technical challenge, and visual appeal. This one was voted Best of Show at the 1976 IPMS/USA Convention in Dallas. (Photo courtesy Arlo Schroeder)

MARTIN PBM-1: This lovely representation of a Mariner in late prewar livery was created by our resident "USN modeller", Arlo Schroeder. (Photo courtesy Arlo Schroeder)

Diego area aviation history, Consolidated/Convair and Ryan Products, U.S. Naval Aviation and World War II Pacific Theatre Operations, as well as a broad coverage of worldwide aviation.

Incidentally, a useful "shoppers guide" for aviation museums the world over is the recently published book *Great Aircraft Collections of the World* (1986) by Bob Ogden, which, aside from text, photos, and listings of museum aircraft, gives the addresses and telephone numbers of 70 major collections, from Finland to Thailand.

BACKGROUND MATERIAL

The underlying motivation for every dedicated modeller is a passion for aviation: for the machines themselves, for the flesh and blood participants in the drama of flight and for the events in which these men, women and machines participated. Another way of putting it is that every serious model builder is also a dedicated aviation enthusiast, usually with his own library of cherished archives. So, while we must beware of straying too far from our subject of scratch building, we will

Chapter II: Hard Core Lore 37

LOCKHEED "Air Express": Flying this aircraft, the irrepressible Roscoe Turner set many inter-city records, culminating in a new East to West coast time of 19 hours, 42 minutes, in May 1930. He was accompanied by his lion cub Gilmore, named after their sponsor, the Gilmore Oil Company, of California. (Rice photo)

BOEING 40B-4: Made by Bob Rice, this 1/24th scale model represents a 40B-4 of Pacific Air Transport, which operated as part of United Air Lines, itself the transport element of United Aircraft and Transport Corporation, a large holding company formed in 1929. All of this fell apart however during the trust-busting activities which followed the October stock market crash. By 1934, Boeing Airplane Company, United Air Lines and United Aircraft Corporation had emerged as separate entities. (Rice photo)

SCRATCH BUILT!

Chapter II: Hard Core Lore 39

touch briefly upon relevant historical aviation material, other than that intended specifically for plans/configuration information. For, after all, each modelling project is usually selected at least in part upon the role which that type, or exact machine played in some activity about which the modeller is keenly interested. This background knowledge leads to judicious selection of colors and markings, as well as configuration particulars – i.e., it assures authenticity of the finished product.

MAGAZINES: Aviation periodicals such as *Flight* and *Aviation* have been around almost as long as the airplane itself. While we've touched upon those journals which have provided plans/configuration information, many others are of value primarily for their coverage of development and operational aspects through text and photos.

In the 1930s, it was the "pulps" such as *Flying Aces* which enflamed youthful enthusiasm for aviation more than any other written material. (Seeing the real thing at air shows, overhead, and at local airports was, of course, the primary stimulus, then and now.) Don't seek them for any hardcore lore though, as the content was almost entirely fictional.

The U.S. magazine *Flying* filled the need for a broadly based journal covering the aviation arena, primarily contemporary. During World War II, it was among the best for popular coverage of those tragic but dramatic times – it included well illustrated features on aircraft, events, technical developments, piloting ("I Learned About Flying From That") as well as occasional futuristic fantasies. (Typical fare included gargantuan commercial flying boats whose wing leading edges were filled with engines, and smiling hubbies departing for work from their garage/hangers in fetching little helicopters. Never depicted was the aerial gridlock which would have been deadlier than the skies over Europe in 1943.) While this fare was grist for our mill at the time, technical/photo information on combat types was, in retrospect, badly outdated until 1944, when security restrictions were eased.

Popular interest magazines naturally abounded during the War. Among the better efforts were *Air News* (usually having the best photo coverage), *Skyways* and *Air Ways*, in addition to Flying. Most petered out thereafter, although *Flying* has, of course, maintained its preeminent status for the aviation minded ever since, and *Air Trails* served the flying modeller for many years.

Britain has always provided grist for the aviation historians mill. The venerable *Flight* was supplemented post-war with such journals as *RAF Flying Review* and *Air Pictorial*. Since 1972, *Aeroplane Monthly* has emerged as the historian magazine "*par excellence*", providing a heady mix of lively, in-depth, authentic articles on topics ranging from aircraft development, first-hand recollection of operations, surveys of aircraft recoveries/restorations, to flying events. *Air Enthusiast Quarterly* (now, simply *Air Enthusiast*) has served its faithful clientele well with features of unsurpassed quality on development/operational history of selected types and on notable aviation events. Their high quality photos, color profiles, and detailed cutaway perspective drawings are especially welcome to the modeller.

Since 1963, Challenge Publications has catered to growing popular interest in historical aviation with *Air Classics, Wings,* and *Airpower*. While often denigrated by serious buffs in their earlier years for their rather sensationalist approach, they greatly improved over the years, often providing articles of lasting value. With publication of a series of drawings by Peter Westburg they earned a place in the archives of serious plans collectors.

MONOGRAPHS: Over the years, a major source of historical lore has been the numerous series of monographs, each covering a specific aircraft type in soft cover booklet format. To our knowledge, this pattern was first set by the Air Age Technical Library, produced by *Model Airplane News* during WWII. A brief technical description and development history was accompanied by general arrangement drawings and photographs. Operational history of the selected US types was basically ignored: after all, it was occurring at the time.

In the mid 1960s, Profile Publications, Ltd., burst on the scene with a wonderful run of small format monographs covering aircraft from the Bleriot to the Concorde; the first 204 in 7" x 9" size with red covers, thereafter 7-3/8" x 9-7/8" with more pages and generally superior color 3-views. Their carefully selected authors were a veritable Who's Who of aviation historian/journalists, including such honored names as Peter Gray, J. M. Bruce, Peter Bowers, Philip J. R. Moyes, Ray Wagner, Bill Larkins, Francis K. Mason, Witold Liss, Gianni Cattaneo, Rene Francillon, Michael Bowyer, J.Frank Dial, Peter M. Grosz, C. F. Andrews, Harry Gann, Alfred Price, Richard Smith, and Martin Windrow. Before the series ran its course, many previously neglected production types had been well described, along with most of the better known military and commercial types through WWII. We only wish they had prevailed for somewhat longer, to have included such categories as WWI German 2-seaters and 1950s jets, both civilian and military. Later, they were reissued in a set of hard-bound volumes. Profiles are fondly recalled and still make a valuable contribution to most collectors files.

In what can perhaps be regarded as an extrapolation of the Profile concept, Squadron/Signal Publications, Inc., of Carrolton, Texas, introduced the ongoing ". . . IN ACTION" series of wide format (11" W x 8-1/4" H), soft cover monographs. These fine, handsome, well researched, yet inexpensive efforts document the development, configuration and operations of each selected type (mostly WWII) through succinct text, many photos, line drawings, and detail sketches, color profiles and dramatic color painting on the front and rear covers.

Another righteous effort in this genre is the ". . . SPECIAL" series, produced by Ian Allen, Ltd., of London. In these excellent hard (slick) cover, yet relatively inexpensive monographs, the format is looser than . . . IN ACTION, with sections

RUMPLER C.IV: Opposite: This 1/32nd scale model by John Alcorn was based upon the MAP plans, heavily supplemented by reference to a French monograph which documented a captured example of the type by text, photos, line drawings, dimensions and performance data. This document was lost in the tragic February 1978 fire which consumed the original San Diego Aerospace Museum. This model received the first Judges Grand Award at the 1977 IPMS/USA Nationals in San Francisco. (Joe Faust photos)

devoted mostly to operations, as appropriate for each type. Each, prepared by a well-credentialled historian/journalist, includes carefully chosen, usually large format photos – often of interior/exterior details to please the modeller, plus a double-truck multi-view color drawing of excellent quality, supplemented with photo-like color renditions.

The "...AT WAR" Series: Although varying considerably in page content from one to another, are full-on hardcover books, describing operations of a given WWII type, primarily through a sequence of personal recollections by aircrew, and occasionally by maintenance personnel. Therefore, the content is lively, with emphasis upon human interest perspectives. For each, the publishers have contracted with a recognized authority, usually one especially knowledgeable of the type under scrutiny. For example, B-17 Fortress was prepared by Roger Freeman. William Hess, the well-known fighter aces historian, prepared the P-47 and A-20 issues: in the B-17 book, he recounted his role as a waist gunner and the September 1944 mission to Blechammer, during which he was shot down and interned.

The Schiffer Military History line includes a growing number of aircraft monographs, most of which are translations of German titles. In addition to the more popular Luftwaffe types (Fw 190, Ju 88, Do 17, He 111, Ju 87, Fw 200, Bf 109 and Bf 110), a welcome number of these photo essays covers such exotica as the Do 335 "Pfiel," WWI airships, Horten Flying Wing, He 100, He 162, He 280, He 219, Me 163, Me 262, V1, V2 rocket, Ar 196, Ar 234, and Lippisch P13a.

BOOKS: Considering the gratifying profusion of books on historical aircraft, we would not attempt to enumerate all of those tomes which we regard as being of primary value to the modeller/researcher. To be sure, the list would be arguable and doubtless some of the "classics" would be overlooked, even if we chose only those books which emphasize the aircraft themselves: configuration, development, and operations. Besides, there are many fine books dealing primarily with the broader contexts in which aircraft were involved (my own favorite in this category happens to be Francis K. Mason's *Battle Over Britain* – a scholarly *tour de force*, if ever there was one – Alcorn).

To be sure, there are good books, long since out of print, going back to the early days of aviation. Janes, of course, immediately comes to mind as a primary source for reliable, comprehensive aircraft information (due allowance being made for security constraints during the two Big Ones). One, for 1919, was of such value for World War I information as to be reproduced in recent years – hard cover, ads and all.

While some hard cover books appeared during World War II, most are of limited value today, nostalgia aside. For example, two *Air News Yearbooks* were produced; photo essays in elongated format. Many of the full page, black and white photos were excellent shots which have become classics for the type. However, in Volume II, we were "ripped off" by substitution of recognition model shots for photos of actual enemy aircraft!

No such cheap shots of course emanated from Janes for 1942 and 1943/44, whose only transgression was airbrush removal of markings from many German A/C photographs. In common with all of the wartime publications until very late in the War, their coverage of Japanese types was spotty at best.

As a result, most of us didn't learn to differentiate the many Japanese types and variants until many years later. (A "Tony" was just something that resembled a Macchi 202 until William Green explained that it was really a Kawasaki Ki-61 "Hein").

Among the first significant books post-war were the Harleyford series from Britain: among other contributions, they were the first since Janes, of 1919, to give a comprehensive presentation of WWI aircraft, including 3-views of each (of limited value to today's sophisticated modeller, but it was a noble beginning).

One day in 1957, while browsing through a bookstore in Hong Kong, I (Alcorn) discovered the newly published *Famous Aircraft Of The Second World War* by William Green, which I instantly purchased (I was a Lt.(jg) aboard a U.S. destroyer). It had an electric effect on me, rekindling an aviation interest rather dormant during my college days, when more urgent distractions prevailed (nevertheless, I had enjoyed watching the many fleet types of the day – F9F Cougars, F2H Banshees, FJ3 & 4 Furys, F7U Cutlasses, and yes Fairey Gannets, from my vantage point on the bridge during plane guard duty). Famous Fighters of course set new standards for historical aircraft journalism, setting the stage for all which has followed since – not the least of which have been the immense contributions by Green himself (and his staff) through the three subsequent volumes of that series, *Warplanes Of The Third Reich* and editorship of *Air Enthusiast*.

Among the many sterling efforts from Albion have been the long Putnam series covering the products of specific British manufacturers: Vickers, Avro, Short, Handley-Page, Supermarine, Hawker, Blackburn, Fairey, DeHavilland, and so on. Each is a fine piece of scholarship, briefly documenting each of the manufacturer's products with text, photo, specifications, and a line drawing. Peter M. Bowers responded with comparable tomes on *Boeing Aircraft Since 1916* and *Curtiss Aircraft 1907-1947*, followed by Rene' Francillon's *McDonnell-Douglas Aircraft since 1920* and *Grumman Aircraft since 1929*, published by Putnam's U.S. associate, the Naval Institute Press.

In a similar vein by Putnam (and their US publishers Doubleday and Funk & Wagnalls) have been superb encyclopedic volumes on *German Aircraft of the First World War*, by Peter Gray and Owen Thetford, *Aircraft of the RAF 1919-1957*, by Thetford, and *British Naval Aircraft Since 1912* by the same author. Rene' Francillon responded with *Japanese Aircraft of the Pacific War* in 1970, setting this fascinating subject right for the first time. The pinnacle, however, was reached by J. M. Bruce with his epic, large format *British Aeroplanes 1914-1918*.

Since 1970, Ray Wagner of San Diego has upheld our cause through *American Combat Planes*, now in its third incarnation.

A new title which should prove popular with modellers is *Mustang: The Racing Thoroughbred*, by Schiffer Publications. Birch Matthews and the late Dustin Carter pooled their knowledge and photo archives to produce a book of surpassing visual excitement, covering the P-51's long domination of the postwar air racing scene, from Cleveland to Reno.

SOCIETY JOURNALS: Every self-respecting (and some less so) facet of collective human endeavor has its society, and every society worth its salt has its journal (and/or news-

Chapter II: Hard Core Lore   41

MARTIN T4M-1: Here are two views of Bob Davies' T4M-1, which shared Judges Grand Award honors at the 1991 IPMS/USA Nationals in St. Louis. (Partain photos)

letter). After all, we have the Society for Prevention of Cruelty to Animals, the Autobahn Society (they spot fast birds driving on Germany highways), and so indeed have we the IPMS, AAHS, AIAA, and *Cross and Cockade* – each with a proper journal of relevance to this survey.

Beginning in 1960 with a hand-typed, stapled cover format, Cross and Cockade grew into a highly respected quarterly journal for World War I buffs. For some years thereafter, enough vets still remained for lively, insightful recollections, usually through interviews with some Cross and Cockader. Remarkable photographic coverage was often obtained, from personal/family files as well as from institutional archives. The organization was active both in the U.S. and Britain, each group producing its own journal.

The American Aviation Historical Society (AAHS) was founded in 1956 by William T. Larkins. Drawing extensively upon material supplied by active members, a dedicated core group from the Los Angeles area has maintained a steady course of quality journalism through many vicissitudes. The quarterly Journal is always choc-a-block with well presented, interesting and thoroughly researched articles covering the full gamut of American aviation (yes, by "American", we mean both the Northern and Southern continents). There is a wealth of inspirational grist for the modeller within the pages of the Journal, especially of those machines outside of the well-documented WWI/WWII categories.

EPILOGUE: Nuthin's forever. There are only a finite number of aircraft photos from the Wright Flyer to the Enola Gay, and a finite – and ever dwindling number of us who reckon that if it doesn't have a propeller, it isn't worth thinking about. So inevitably must the spotlight of aviation interest move forward into the 1960s, 1970s, and 1980s. But that's for a later generation to record in words and plastic.

GOING IT ALONE: SCRATCH BUILT PLANS

By now, most of the more important and many lesser known aircraft produced through the 1950s have been delineated in reasonable fashion by commercial sources, although many worthy efforts have long been out of print.

There remain those aircraft which, through obscurity, loss of records, or simply fate, have never been adequately delineated in scale – at least in any easily retrievable source. Often enough, such machines attract the scratch builder, for whom "scratch" takes on an added dimension: he must dig for even the source material from which to prepare his own drawings. For such arcane endeavors, source material may include factory drawings, specifications, photos, technical documents, such as Erection and Maintenance Manuals produced for production military aircraft, or similar material featured in long out-of-print aviation journals.

Sometimes one can, or must, avail himself of taking measurements and detail photos of actual aircraft, perhaps through the indulgence of some museum.

But sometimes, as in the case of certain "Golden Era" air racers, nothing exists, short of a few good photos, plus an all too brief list of basic dimensions. A few personal case histories can illustrate the extremes to which the dedicated (fanatical?) modeller can go to create the object of his desire from "almost nuthin."

CASE HISTORY #1: Douglas A-20 (Alcorn): I have harbored an abiding fascination with this aircraft ever since its heyday, during which time I fashioned several rather primitive examples in wood from the 1/72nd recognition model plans of the time. Since then, the A-20 has fallen into relative obscurity, despite its yeoman service with the French, RAF, USAAF, SAAF, RAAF, and Soviets.

In 1970, I set out on a crusade to help rectify this deplorable situation, by documenting its story, preparing authentic scale drawings and building a 1/32nd scale model therefrom. Easier conceived than achieved.

I had already prepared a three-part series of articles for the AAHS Journal on the Wartime Exploits of the Third Attack Group of the 5th Air Force, operational in New Guinea and, later, the Philippines. Their 89th Squadron operated strafer-modified A-20As between September 1942 and November 1943; thereafter the entire group converted to A-20Gs (Part 3 of this series never appeared).

Existing configuration material in my files included 1/48th scale 3-view drawings from an old Maircraft kit (not very detailed but apparently pretty good in general arrangement), a March 1960 MAN feature by Willis Nye (very detailed, but grievously flawed in many important respects) and an extensive 1944 article in Aviation which included a 3-view general arrangement drawing with certain basic overall dimensions.

Realizing at this point that I needed professional help, I wrote to Doug Pirus of AAHS – and Douglas. He kindly responded with a long, impassioned letter explaining that it was next to impossible to extract reliable configuration information from corporate files. This had the curious effect of transforming my academic interest into a burning obsession. Having recently joined AAHS, I soon attended a meeting where I rather timidly approached Harry Gann – another AAHS stalwart, and noted expert on Douglas aircraft. He sent some Xeroxed Maintenance Manual material, including a moderately dimensioned 3-view of an A-20B. More Erection and Maintenance Manual material was soon to hand from various sources, at which point I began preparing 3-views of my own. Numerous omissions, perplexes and discrepancies soon were revealed.

On a tip, I then made contact with the good folks of the Antelope Valley Aero Museum in Lancaster, California (Lee Embree, et. al.) – owners of the ex-Howard Hughes A-20G/VIP conversion (fortunately, not very radical). Through the kind indulgence of Paul Price, "crew-chief" of their A-20, I measured this A/C inside and out during the course of three separate trips from Palo Alto. Each time I returned home with many dimensions and photos, only to experience the eternal dilemma of the researcher – the more I knew, the more I knew I didn't know and the more determined I became to sort it all out.

Sometime during the course of this effort, I began work on my 1/32nd scale, scratch-built A-20A – the first plastic model I had ever attempted, kit or otherwise. Two years and 2400 hours later it was completed, just in time to enter in the 1974 IPMS/USA Nationals, held that year in Anaheim (near Disneyland). To my abject amazement, it won Best Of Show. While the trophy was mine, much of the credit was due George

Chapter II: Hard Core Lore

This publicity photo was taken at the 1968 British IPMS Championships in London. Harry Woodman (left) is holding the National Trophy Cup, awarded for his 1/48th scale Voisin. At right is Jack Bruce, the distinguished aviation historian and chief judge for the event. Fred Henderson, then IPMS President, is in the center. (Harry Woodman)

Lee, my colleague at SLAC and modelling guru, who had initiated me into the Oriental mysteries of vacuforming, air brushing and silk screen decal making, all while under the heady influence of MEK vapors. He didn't actually do any of the work, of course, and my years of solid modelling was a good background, but his advice and inspiration sustained me throughout.

During this period and after, material from Harry Gann and others had grown to a veritable flood of Hard Core Lore, including factory detail photos, data, and a series of general arrangement drawings providing many further important dimensions not given in the E & MM fare or achievable by my rather limited means of A/C measurement (yardstick and tape measure). Eventually I had adequate dimensional/configuration/internal equipment material for preparation of an accurate, detailed set of exterior and interior multi-sheet drawings of the A-20A(& C)/Boston III (& IIIA) bomber and A-20G variants.

Having thus labored so hard and long on the drawings, I realized that it would be at once satisfying and useful to get them into print. So, it was with a great sense of pleasure on my part that they were accepted by *Scale Models* as the basis for what emerged as MAP Plan Pack No. 2989, by no less a draughtsman than Arthur Bentley. (They were featured in the January 1978 *Scale Models*.) He enhanced my material with views of the DB-7, Havoc and Turbinlite variants. My only misgivings regarding the feature was a distressing misquote in the opening sentence of the accompanying article, to the effect that: "The A-20 drawings were prepared by the author based upon relatively limited information collected by that time (summer 1972) in order to construct a large scale model."

Although I don't recall exactly what I did say, the intent obviously was that, while the model was begun with relatively limited information, the featured plans were based upon extensive research and data obtained subsequently. Hope that didn't put off too many potential users of the A-20 drawings.

This success led directly to preparation of an extensive two-part series for Historical Aviation Album covering development, production, configuration and operation of the DB-7/A-20/Boston series – accompanied by Paul Matt's drawings of the A-20G, based upon my originals. With publication of these features (In HAA, Volumes XV and XVI) I felt that my mission to help save the A-20 from obscurity was accomplished – with one exception: before I lay down my Optivisor and Swiss files for the last time, I intend to construct a 1/32nd scale Boston III of No. 88 Squadron RAF; as a companion piece to my 3rd Attack Group A-20A and as my own tribute to James Pelly-Fry and his brave comrades of those distant days.

Incidentally, *Fighters of World War II*, Volume 2, of the recent Argus Aircraft Archive series includes a repeat of the MAP Plan Pack on the A-20/Boston – fortunately without that dreadful lead-in sentence.

CASE HISTORY #2: Rumpler CIV (Alcorn): I include this partly as an excuse to recount the dramatic circumstances surrounding the seemingly mundane activity of returning a borrowed document from museum archives.

Sometime in 1976, I decided to model the Rumpler CIV, an aircraft that I had long admired for its elegant proportions, excellent high altitude performance and yeoman service over the Western Front. During the course of my search for configuration documentation, I visited the San Diego Aerospace Museum. There I discovered an original French report describing a captured example in extensive detail; text, photos and well dimensioned drawings. I borrowed the report, which became the primary basis for my 1/32nd scale effort, along with Ian Stairs' MAP Plan Pack No. 2963.

On 22 February 1978, I called Bruce Reynolds, the museum archivist, to arrange for return of the report that evening. A retired engineer, Bruce lived in the museum during the week, commuting home to Santa Barbara on the weekends. At about 7:00 p.m., I knocked at the back door. He let me and my 13 year old son, Stewart, in and we retired to the archives room to talk airplanes for awhile. About twenty minutes later, Bruce smelled smoke, so we entered the main exhibit hall to investigate. We soon discovered fire trucks outside, attending to a blaze in the adjoining aviation Hall of Fame exhibit.

Since it appeared that all was under control, we returned to the archive room, occasionally returning to the main hall for assurance that all was well.

Perhaps 45 minutes after we had arrived, the archive room was plunged into darkness, presumably by fireman turning off the electricity. We floundered for an emergency lantern, exited through the back door and looked back at a sight which we would never forget – the entire roof of the museum was afire!

Shortly thereafter, flames penetrated the main hall and were pouring through the door. I had the miserable experience of watching their priceless exhibits being immolated. Somewhat later, I entered the north wing under the pretext of being

SCRATCH BUILT!

a volunteer firefighter. We entered the darkened hall, separated from the main area by a full height wall, where among other things, a freshly restored Curtiss JN-4 "Jenny" proudly reposed. Soon the flames began to penetrate the wall, whereupon we beat a hasty retreat. As I departed, I instinctively grabbed a flying type scale model sitting on a storage shelf – it was a Guillow "half inch scale" Rumpler CIV! (which I returned to the Museum in 1987). Thereafter, we watched as much of the awesome spectacle as we could bear, sobbing bitter tears.

They have a fine new Museum there now, worthy of San Diego's historic role in aviation. The aircraft exhibits are splendid and the extensive archives are in the capable hands of Ray Wagner, a renowned aviation historian. But sometimes I'm still haunted by memories of that Rabaul Zero, Ryan M-1, Consolidated PT-3, the Flying Tigers artifacts, the dollar bill that Lindbergh had thumb-tacked to the dashboard of the Spirit for luck, that lost Rumpler report (Fortunately, I had xeroxed the complete document and photocopied many of the illustrations) – and those hideous, all-consuming flames.

Postscript: The blaze had been started by vandals, who had set fire to trash within an opening in the "temporary" frame structure, built for the 1915 Panama-California Exposition. The flames soon progressed upwards into the roof, gradually moving laterally into the Aerospace portion.

While many priceless treasures were lost, the subsequent response from benefactors has been tremendous, resulting in the acquisition of many fine aircraft, artifacts, and archives. They are now housed in a far safer, permanent structure. That it is once again a world class museum is due in no small measure to the professional dedication of its staff and volunteers.

MODELLING TECHNIQUES

Over the years there have been many magazine articles, monographs, and books describing modelling techniques of relevance to the scratch builder.

MAGAZINES: Before and during the war years, *Air Trails* and *Model Airplane News* occasionally carried features on solid model building. The tradition was carried into the plastic era by *Aeromodeller*, *Scale Models*, and *Airfix Magazine*.

Since 1983, the U.S. publication, *Fine Scale Modeller*, has consistently featured excellent, in depth articles on a wide range of modelling techniques, a number of which are referred to throughout this book.

Another US magazine, *Scale Modeler*, also runs "how to" features. Bob Rice's September 1987 article "Golden Age" airliner, describing construction of his 1/32nd scale Boeing 80A-1, is especially valuable to scratch builders.

IPMS JOURNALS: These publications (USA, Britain, Canada, Australia, etc.) have run countless "how to" articles, covering specific subjects as well as basic modelling techniques.

MONOGRAPHS: Kalmbach Books, the publisher of *Fine Scale Modeler*, has also produced monographs on our hobby, including *Building Plastic Models*, *Hints and Tips for Plastic Modeling* and *How To Build Plastic Aircraft Models*.

BOOKS: We have already mentioned James Hay Steven's, *Scale Model Aircraft*, published in 1933. Then, in 1975, Harry Woodman acknowledged his debt to Stevens in the dedication of his book by the same name, with ". . . In Plastic Card" appended in smaller letters. We hope that our book is worthy of this tradition.

This 1/16th scale George Lee representation of the 1924 Pulitzer Trophy winning Verville-Sperry R-3 resides in the NASM. (Joe Faust photo)

CHAPTER III: VACUFORMING

Preparation of vacuformed fuselage, wing and various smaller components is the very essence of scratch building, yet this crucial stage represents but a small fraction of the total "manhours" required for the project: and the vacuforming operation itself requires but a few hours at the most; far more effort and skill is required to prepare the forms.

CARVING THE FORMS

For us old solid modellers, carving the fuselage, wing, nacelle, cockpit, etc., forms from hardwood is very satisfying, taking us back to our modelling roots. In other words, the first actual construction step is to build a solid model. It often looms as the most daunting hurdle to anyone lacking such background. For this reason, it is worth dwelling upon the specifics in some detail.

SIKORSKY S39B: This lovely 1/32nd scale model by Bill Bosworth depicts a Varney Air-Ferries machine which flew commuter flights between San Francisco and Oakland in 1932 – rapid transit before the Bay Bridge and BART. Bill described construction as follows: "The hull/fuselage is a combination of vacuform (top) and traditional boat construction (built up frame and skin). The wing is solid plastic! I have absolutely no idea why I did this. Vacuforming would have been much lighter and easier. The markings are hand painted." (Bosworth photo)

SCRATCH BUILT!

A carefully contrived study, showing the profile contoured white pine fuselage of a 1/24th scale Bf109E, based upon Doug Carrick's MAP plans. All of the work to this point was done with knife, wood file and small carpenter's square. (Alcorn photo)

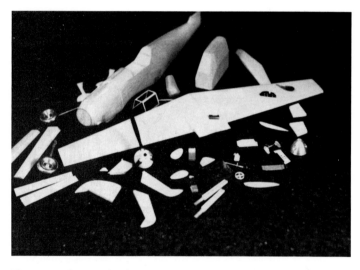

Here are shown all of the basic components for my 1/24th scale solid Bf109E, circa 1970. The metal wheels and spinner were turned by a machinist friend. (Alcorn photo)

MATERIAL: The most commonly used form material is basswood (from the Linden tree). It is hard enough to permit inclusion of complex, crisp detail without being unpleasantly difficult to carve. Its best characteristics are a very even texture with minimal grain, and lack of sap in properly cured specimens – in contrast to, say, pine. Oak can also be o.k., though it's awfully hard.

FUSELAGE OUTLINE: Let's use the example of a 1/32nd scale Douglas Boston III, based upon my (Alcorn) Scale Models drawings. The 18" long fuselage side view is traced (on a light table or against a window) on stiff manila paper, which is then carefully cut out using a sharp #11 Xacto or surgeon's blade. Using this template, the side view is then marked upon the mating surfaces of two good quality basswood blocks – each of an appropriate plan view thickness to form one side of the fuselage.

The side view of each half is then rough cut out, using a knife or jigsaw. If a knife is used, the perpendicularity of any given cross-section can be maintained by periodic checking with a small carpenter's square. The side contours are brought into final shape using wood "bastard" files, and finally, sandpaper blocks. At some point in this process, allowance must be made for the as-formed thickness of the polystyrene shells: that is, the fuselage contour must be appropriately undersize all around (use 30 mils in the case of 40 mil sheet). This can perhaps best be achieved by drawing and cutting the heavy paper template with this allowance. Incidentally, marking the side view on both sides of each block – registered by carefully scribed lines across the front, back and top – will facilitate the process of achieving registry across the faces.

Now, the two fuselage halves are lightly glued together by running a bead of wood glue along their mating edges. (Don't glue over the entire face, or you'll later have great difficulty separating the halves for the molding process.) Once glued, the top and bottom surfaces should be given a final truing with a sandpaper block. This ensures that the two halves are an exact match and that perpendicularity of the edges to the sides is maximized.

At this point, the top (and bottom) views are marked on the block, again using heavy paper templates – registered to the joint between the two halves and to fore and aft marks or edges. Sure, the side view contours mean that fore and aft repositioning of the template will be required as marking progresses. Now rough out the top view with a knife; followed by filing and block sanding to bring in the top/bottom contour. Again, make allowance for thickness of vacuformed shell, so that the final product won't be "fat" by that amount.

An option to the above is the technique described in the wartime recognition model instructions, employing a jigsaw. In this case, lightly glue the two fuselage blocks together: here, a wood glue pattern across all of the mating surfaces may be appropriate to prevent the blocks from later falling apart. Scribe the side view and carefully bandsaw just outside the lines. Reattach the fragments by light gluing. Then, scribe and

Air brush and compressor aside, this was about it for my Wedell-Williams #44 models. The bottle is just about out of MEK: the slide rule is *de rigour* for scaling from photographs, etc. (Alcorn photo)

Chapter III: Vacuforming

saw the top view. Now, remove the remaining upper and lower side view fragments by brushing solvent (lacquer thinner or MEK) into the seams and prying lightly with a knife. Side and top contours can be trued-up using file and sanding block, as before.

CONTOURING: This is the most satisfying portion of the entire form-making process, as well as the most demanding. But, George insists that it's (Peking) duck soup: just cut away everything that doesn't look like an A-20.

First, "female" cross-section templates are cut from heavy paper, having first traced them from the plans on a light table. Again, remember to make allowance for thickness of the plastic sheet. For the A-20, there are eight cross-section locations, which must be defined by some appropriate means. One option is to separate the fuselage halves by the solvent and prying method described above. Mark the cross-section locations from the side view template and then carefully reglue the halves. Take this opportunity to give a final check of the side view contour – typically, a little truing here and there will be appropriate.

Now, unsheath the old carving knife and begin whacking away. Pretty soon you'll have wood chips in your hair and coffee, as well as over a large portion of the floor. Despite all precautions, some fraction of this mess will find its way into all rooms of the house – to the great consternation of mom, wife, girlfriend, or roommate. No matter though, they love you despite your peccadilloes, right? At any rate, this is a good test. Occasionally, a telltale chip will appear in odd places for years to come.

At first, caveman tactics are appropriate, since there is lots of material to remove. Incidentally, the time-honored Xacto whittler's blade in the fat holder is fine, although at first you may prefer to use some more robust weapon, as long as it's sharp.

Generally, the cuts should be fore and aft, shallow and elongated: this is efficient and minimizes the chance of accidentally removing a divot from the mold volume lurking beneath.

Woodforms for Wedell-Williams #44. Well, half of them anyhow. Funnily (as some Brits say), Bob Rice used these forms many years later for a movie promo model. (I had demured, being off on some other project at the time). (Alcorn photo)

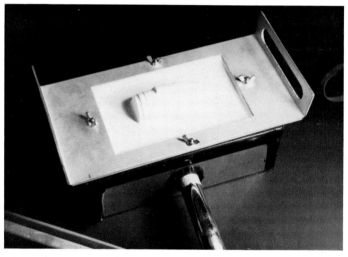

Here is the basic Lee/Alcorn vacuform rig, showing vacuum box (with vacuum cleaner tube inserted), and aluminum platen. An inboard A-20A nacelle half has just been formed. (Alcorn photo)

These are the basic hardwood masters from which the shells for (one half) of my 1/32nd A20A were vacuformed. Note that the fuselage half consists of three sections: nose, main portion and tail. A separate form is used for the pilots and gunner's plexiglass canopies. (Alcorn photo)

Eventually, you'll have removed enough material that it will be appropriate to begin test-fitting the templates. Initially, you'll be quite a way off – the cross-section will still be "too square." Now, it's time to sheath the assault weapons and abandon the power grip (the Xacto whittler's blade may still be appropriate, though possibly now held in a smaller handle). Free whittling cuts have given way to shallow slices, with the blade guided by the thumb of the piece-holding hand. You're now making the transition from wood butcher to craftsman: face distorting grimaces and low growls have given way to thoughtful lip chewing. As a symbolic gesture, you may even be moved to sweep the floor, but don't bother vacuuming.

By now, you're carefully pecking and checking. At some point, the wood file will begin to supplement and then replace the knife as your basic tool. It, of course, has the great

SCRATCH BUILT!

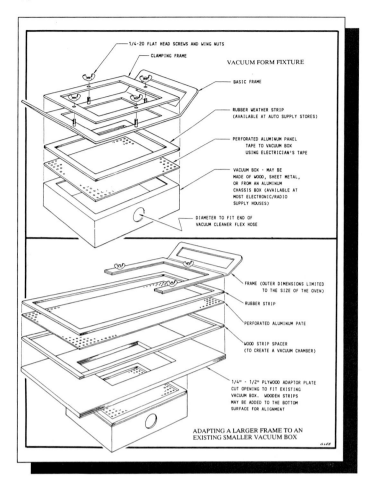

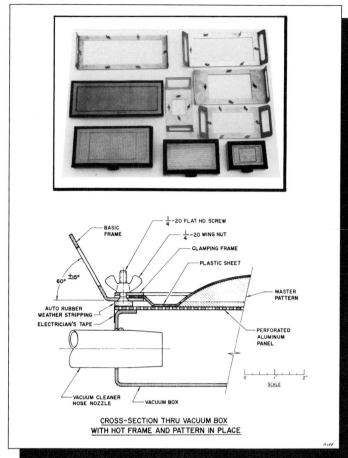

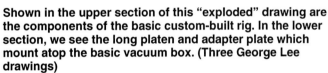

Shown in the upper section of this "exploded" drawing are the components of the basic custom-built rig. In the lower section, we see the long platen and adapter plate which mount atop the basic vacuum box. (Three George Lee drawings)

advantage of generating smooth transitions from one cross-section contour to the next, and of replacing the chiseled surface with one that is "rough machined." Now, aside from nose contours, about which more later, the A-20 fuselage is straight forward. Basically, one fuselage section contour blends into the next, with no concave regions.

Two contrasting examples which come to mind are the fuselages of the Dornier Do-17Z and the Douglas F4D Skyray, both of which have extensive fuselage to wing transition regions — so much so that it would be appropriate to split the form along the horizontal rather than vertical plane for subsequent vacuforming. To "bring in" concave surfaces, the curved backside of a half-rounded wood file is invaluable, as well as various sizes of rat-tail files. For such exotic shapes, one should refer constantly to detail photos, since cross-section templates can't tell the whole story.

Finally, even the file and occasional knife sliver will give way to sandpaper: folded, rolled or block-mounted as appropriate. As you proceed to finer grades of sandpaper, you're entering the final phases of form preparation. Inevitably, during this tedious and time consuming phase, you'll occasionally revert to file or knife and – yes – despite all precautions to the contrary, to application of a dab or two of "plastic wood."

In the case of the A-20, and of many other aircraft, it is now appropriate to perform major yet delicate surgery upon your fuselage form — namely, removal of the nose region. Such amputation is necessary for one or both of the following reasons: to provide a separate form for molding the nose section in clear plastic and/or simply to divide the fuselage into sections which can be accommodated by your vacuforming rig. This can be accomplished with an Xacto (or equivalent) saw blade. The crucial thing here is to accurately mark the cross-section to be cut on the form surface — it must be perpendicular to the fore and aft centerline and to the parting plane.

FORMS FOR SECONDARY FEATURES: Engine nacelles, tail booms as for a P-38, or large floats are made just like fuselages. Certain shallow bumps, lumps and gentle surface anomalies are best vacuformed integral with the primary element. Such features can be produced in one of two ways on the wood form: by carving integral with it, or by carving separately and gluing onto the larger element (with the help of glue or plastic wood edge filleting, as appropriate). For large lumps and/or those with angular cross-section and main surface intersections, it is best to carve a separate form, vacuform the item and fit it in place later. Below some critical size, just carve the item in plastic and MEK it on – or, if angular, build it up from plastic sheet. Vacuform fabrication of certain specialized features will be discussed later.

Chapter III: Vacuforming   49

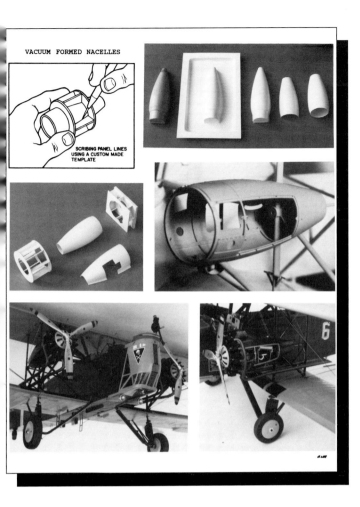

MOUNTING THE FORMS

First, the completed fuselage forms must be separated, by solvent painted into the seams to loosen the glue, followed by running a knife blade into and along the joint to break it apart – if the knife blade breaks instead, you'll know that you used too much glue.

In order that the vacuformed sheet plastic pulls down clear to the centerline of the wood form, it must be mounted upon a base of suitable height – 1/2" is about right for large forms such as a fuselage. Balsa is an adequate and convenient material for the base, which must be carefully trimmed and lightly block-sanded to be flush with the edge of the hardwood form. Once done, a scribing tool can be run around the seam, indenting a "parting line" into the balsa: this leaves a convenient centerline trim mark on the polystyrene shell.

No vertical "cliffs" should be left exposed on the hardwood form, such as that created by the severed nose portion. So, a balsa block should be glued in place and carefully faired into the form. In this manner, the styrene sheet won't be pulled thin as it passes across the end of the form. This feature is shown in the accompanying photo, illustrating the mounted fuselage form.

Incidentally, even though you may never need to remove it, it's a good idea not to glue the base too extensively to the form – you may later want to reassemble the form segments for some reason.

Over extensive form surfaces, especially where concavities exist, it may be advisable to drill several 1/32nd diameter holes into the form and clear through the base, in order that good vacuum is experienced over the entire forming surface of the polystyrene sheet.

THE VACUFORMING RIG

The archetypal personal vacuform rig was that produced by Mattel in the 1960s. It was a self-contained device, embodying the platen, heater, and vacuum box features. Unfortunately, it was of modest size and worse, is no longer produced, being something of a collector's item by now.

We have always used a "custom-built" model, whose plans are shown here. Making it is no problem if your workshop includes a Bridgeport Mill – otherwise, it helps to have a machinist-friend with access to such a device. The rig is really quite simple, consisting of a vacuum box, a set of various sized platens, and an adapter plate to accommodate the larger ones.

The vacuum box is simply an electronic can, or equivalent: modified by replacing the top with a perforated metal cover and a sidewall hole to receive a vacuum fitting. Perforated 1/16" thick plate is cut to size and matching holes drilled in the box top flange and plate. It is attached by flush screws, with a thin sealing gasket trapped between plate and box flange.

For "system vacuum" we use the family tank type cleaner (Electrolux, Eureka, or equivalent), with its hose snout inserted through the box vacuum fitting.

The platens are where the Bridgeport comes in, to machine the frames from plate stock. Either structural aluminum (2024, 6061, 5083, etc.) or 300 series stainless steel will do. Our machinist-friend bent up the ends of the upper platen elements after machining holes therein, to form handles. But, separate handles could of course be made and screwed on. A crucial element for proper function is a series of studs attached to the lower platen and projecting through matching clearance holes in the upper plate. The studs can be attached in any of several ways, the important points being: firm attachment so they don't pull out; reasonable vacuum tightness; correct alignment; and bottom ends flush with the frame. For the resourceful modeller lacking either a Bridgeport or a machinist-friend, other platen options are surely available – after all, the requirement is simple enough.

VACUFORMING TECHNIQUE

For demonstration purposes, we'll form a fuselage half for the 1/32nd scale A-20. Since the form, including end blocks, is 12 inches long, we must use the 18" long by 7" wide platen and adapter plate.

For a large component such as this, 40 mil ("thou") high impact polystyrene sheet stock should be used. With a mat cutting knife and long steel straight edge, cut out from commercially available 48" x 48" sheets several pieces which just fit within the studs of the lower platen – 6" x 16-1/2", in our case. Place one such piece between the two platen elements and dog them together using the wing nuts.

Fire up the kitchen oven to 250-275 degrees F: hopefully, it has a window and inside light. Set the vacuform rig nearby on a stool or counter. Plug in the vacuum cleaner and place the form on the rig. Have a good, clean pair of heavy cotton workgloves at the ready. Now, put the loaded platen in the oven, on a rack, with the handles down so that the drooping plastic can't touch the rack. Watch it closely until the plastic has sagged 1/2 inch or more. Turn on the vacuum cleaner and (with gloves on) quickly open the oven, grab the platen, turn it over and bring it down smartly and squarely upon the rig. Yreka! You have an A-20 fuselage half – or . . .

- the platen wasn't brought down squarely upon the vacuum box (or its adapter plate), so that the wood form is too near one edge of the plastic. Further, if the mismatch is too great, vacuum may be lost due to a gap between platen and seal. Mismatch is a common error, which simply requires practice to avoid;

- the plastic didn't pull down past the centerline (base) of the hardwood form. If the plastic was hot enough, no time was wasted between oven removal and placement, platen registration over the vacuum box (or adapter plate) was good, and full vacuum (suction) was obtained, then the fillet radius at the base of the 40 mil plastic should be about 1/4 inch. A significantly larger fillet radius usually means inadequate vacuum, which usually results from leakage through the seals, or through seams in the vacuum box;

- you tripped over the vacuum cleaner cord.

– *Errare Humanum Est*

Incidentally, the plastic from a faulty pull can usually be reused at least once. Simply put the platen/plastic back in the oven, making sure that the shaped plastic doesn't touch the rack. As it heats, the plastic will recover its unformed (but droopy) shape, as if by magic.

Once a good shell has been obtained, it's an easy matter to trim off the excess sheet – so long as the form-to-base seam line is visible on the inside surface.

CANOPIES

(CANOWORMS may be more descriptive, in this context.) In a very real sense, this is where the gentle art of vacuforming plastic model airplane components began. Certainly, they were the first such items to appear in kits.

Back in the mid-1940s, someone came out with a solvent which was supposed to soften clear acetate enough for the state-of-the-art modeller to pull down over his carved form, thus becoming the first in his precinct to exhibit a P-51D with a blown canopy. We usually blew it alright – saving the model by the simple expedient of grafting the form onto the fuselage – and then painting it with the canonical silver/grey.

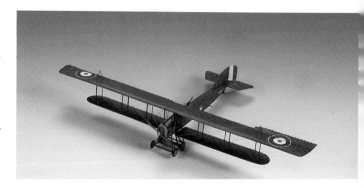

SHORT BOMBER: This ungainly affair was Britain's first serious attempt at producing a strategic bomber. Eighty-three were built in 1916 for the Royal Naval Air Service: all but the fifteen Sunbeam-built examples were powered by the 250 hp Rolls Royce Eagle. Clearly developed from the Short 184 seaplane, the Bomber featured wings of enormous span (85 feet) for a two-seat, single-engine aircraft: the fuselages of most were lengthened from the 184 for directional stability. The type's appearance, challenge and obscurity attracted George, who based his 1/72nd scale model on the Harleyford drawings. (George Lee photo)

Then, of course, you could always heat the plastic sheet in boiling water, usually producing a bubble on the thumb rather than in the plastic. Eventually, Mattel came to the rescue, by which time acceptable canopies (by the standards of the time) were appearing in kits.

I (Alcorn) have tended to vacuform canopies for larger scale efforts from the real thing – Plexiglass (Lucite, or Perspex if you prefer) sheet. Its advantage is that it is very strong and hard enough that its surface can be worked after forming and then polished back to high optical clarity. Its disadvantage is that it is very hard to cut – it must be sawed and the edges then sanded to final shape. Also, I've never found sheet thinner than 1/32", thus limiting its use to larger scale models. Plexiglass must be oven heated to around 300 degrees F.

After vacuforming, the inner surface can (and usually must, in my experience) be sanded to eliminate surface blemishes: you can work down from #400 grit, or rougher if necessary, eventually converting to aluminum oxide rouge and finally to automobile polish.

Peter Cooke typically uses optical grade 10 thou PVC sheet, heat formed over a shape whose surface is also "optical grade", so that nothing beyond fine polish is later required. But, for canopies, turrets, etc., having extensive framing (such as his Lancasters), he casts in clear epoxy, using techniques described in Chapter V.

Incidentally, two useful articles on modelling canopies appeared in *Fine Scale Modeller*: August 1986, "Vacuum Forming Canopies", and February 1988, "How To Install Canopies and Clear Parts."

CHAPTER IV: THE BASIC MODEL STRUCTURE

INTRODUCTION

The internal structure of a scratch built static scale model serves three basic roles: it supports the thin vacuformed shells (or sheet skin) against gross and local distortion; provides hard points for component assembly; and simulates visible structural elements of the actual aircraft. This vital and time consuming structure constitutes a fundamental difference between scratch built vacuformed and commercial injection molded models, since the latter have thicker skin and integral structural features.

While almost all scratch built plastic models require some vacuformed components, the extent ranges from a few cowling panels on many WWI machines to shells for all of the major elements of metal specimens with double curved surfaces. In many respects, constructing a plastic static model of early

IL'YA MUROMETS G-3: This is Bob Rice's magnificent Il'ya, which now reposes in the NASM. It won Popular Best of Show at the 1989 IPMS/USA Nationals in San Diego, as well as First Place in the large scratch-built aircraft category. Bob's 1/32nd scale model was constructed from drawings prepared by Harry Woodman, probably the world's leading authority on the type. Both Rice and Woodman were invited to – and attended – the 100th anniversary commemoration of Igor Sikorsky's birth, held at Kiev in May 1989. An excellent article on construction of this model, including Woodman's drawings, is provided in Fine Scale Modeller, March 1991. The aircraft represented by Rice's model was constructed in 1916 by Russko-Bal'tisky Vagonny Zavod of St. Petersburg. Some 50-odd Il'ya's were built, enjoying a long service career during WWI flying long-range reconnaissance and bombing missions – mostly with impunity. The first successful four-engined bomber, it was a remarkable achievement for Czarist Russia, whose other aircraft were undistinguished. (Rice photo)

SCRATCH BUILT!

These are the main components of my 1/32nd scale A-20A: *Deo Volente*, a similar Boston III will materialize, in the fullness of time. (Alcorn photo)

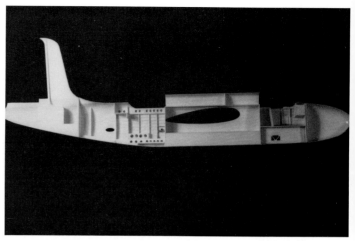

This shot of the interior half of my A-20 shows the model structure, the visible interior in the pilots and gunner's compartments, and the wing cutout. (Alcorn photo)

fabric-covered machines is not unlike building a "flying model" in the classic manner. Consider for example, building a 1/32nd scale Keystone. While many engine and cockpit area panels must be vacuformed, the fuselage structure is constructed using plastic "stick stock" for longerons, and truss bracing and sheet stock for bulkheads. As discussed below, wing structure would employ sheet stock ribs mounted upon brass tubing spars and stick stock leading edges. Assembly and rigging technique would, of course, have distinct flying model affinities.

FUSELAGE

VACUFORMED SHELLS: Using the A-20 as an example, the structural work begins when the shells have been trimmed to their centerline marks. This aircraft was of monocoque construction: in any case, interior features are visible only in the crews compartments. In non-visible regions, it is only necessary to place sheet stock bulkheads for rigidity; and wing/tail assembly attachment features.

Our preference is to build up fuselage halves for later centerline assembly: in this case, half bulkheads are installed – about 1-1/2" apart. In the visible crew compartment areas, bulkheads, formers, stringers, decks and such which comprise the actual structure must be simulated. First though, smooth the inside surface of the shell with sandpaper, to remove any evidence of the form. Where light fore and aft stringers are present, they should be added first, with appropriate slots cut in the intersecting formers and bulkheads. In constant cross sections where the stringers are parallel, they may be trued by running a slotted former back and forth before the MEK-attached stringers have set permanently to the shell. At about this point, with perhaps a temporary bulkhead or two for rigidity, each knife-trimmed shell can be trued over its centerline parting plane by gently dragging it back and forth across a big sheet of sandpaper (say 180 grit) laid upon a flat surface. While taste and circumstances may vary, generally the canopy and other areas later to be cut away should be left in place until the shell has been well stiffened by its internal structure, but before proceeding beyond this point.

In crew compartments especially, many secondary features should be added integral with the shell structure. These include brackets and shelves for equipment and perhaps even a "black box" or two. Then after careful cleanup with knife edge scraping, fine sandpaper and the like, the visible interior is airbrush-painted in the appropriate color.

Wing/fuselage attachment demands careful planning and execution in order to ensure accuracy and strength of fit. Location, dihedral, angle of incidence, plan and transverse alignment, as well as strength and method of attachment are factors which must all be considered before adding the fuselage features. The technique to be employed will depend upon specifics of the aircraft as well as personal preference.

In certain cases, it is appropriate to terminate the wings at their root interfaces, perhaps as on the real thing. Tubular brass sections could be built into the fuselage to receive wing spar extensions of the next smaller size. For wings with heavy, rectangular spar extensions, sockets can be constructed within the fuselage center section. Some jigging and fixturing may be required to properly align these features.

The basic alternative to wing root joints is to provide a well stiffened cutout for later insertion of the complete wing. Such might well be the best choice for low winged aircraft such as Bf 109 or P-51 (a Spitfire though is probably best served by wing root attachment, due to its extensive fillets). However, shoulder and mid-wing aircraft can also be assembled in this manner. I (Alcorn) chose this method for my 1/32nd scale Douglas A-20A. I accurately cut out the wing/fuselage junction airfoil on the two shells and then reinforced the area. Since the tapered wing was larger at the centerline, I assembled the two fuselage halves over the wing. The cutouts automatically produced correct wing positioning in all but plan view perpendicularity. A wing with constant central region cross section can, of course, be inserted through cutouts in the assembled fuselage.

Full wing assembly for mid-taper-wing aircraft can also be achieved by the solid modeller's time-honored method of removing the fuselage section above (or below) the wing and later reinstalling it with appropriate seam filling.

Chapter IV: The Basic Model Structure

Before joining the fuselage halves, any interior details which cannot be conveniently reached later should be added. This can be frustrating since, at this point, the emotional sentiment is to get on with major assembly. Just about any feature attached to the sidewalls falls into this category, including control units and myriad "black boxes." Control cable runs should also be added.

Transverse elements which cross the centerline demand some premeditation. These include visible bulkheads, decks and instrument panels. Despite your best placement efforts, you're lucky if preattached bulkhead halves match exactly at assembly – even if they do, there's still the seam to contend with. Nevertheless, this is usually what must be done. Seam matching and joining is facilitated by placement of an overlapping, full height strip behind one of the halves. Unless they are badly mismatched, this will provide alignment and a surface for glueing – by the simple expedient of painting the visible seam with MEK and letting it capillary onto the strip behind.

SLAB SIDES: For the slab-sided fuselages of most older aircraft, vacuforming is inappropriate – except for certain double-curved coamings and turtledecks. Whether the flat or single-curved surfaces of the original aircraft were plywood or fabric covered, the "box" method is appropriate for its simulation. Exemplars featured herein include Rice's IL'YA, George's Keystone and my Rumpler.

The sides and bottom are cut from (typically 30 thou) sheet polystyrene. Where interior structure of rectangular section wood is to be visible inside, this is represented by styrene strips of the appropriate size, glued to the side and bottom panels prior to fuselage assembly. A typical example is that of my DH9A – see accompanying photo on page 54.

The box structure is assembled using solid rectangular bulkheads where representative or not later visible – and transverse styrene columns. The trick of course is to achieve symmetry in plan view, as the rear portion of the sides converge to meet at the rudder post. The problem is reminiscent of that encountered in our youth during construction of stick and paper fuselages for flying models. In our case, however, the process is simplified if the underside can also be sheet, precut to the plan view contour, though undersize to accommodate full depth side profiles. A few judiciously placed transverse bulkheads can be first attached to the underside slab, in order to facilitate subsequent side slab attachment/ alignment.

As with vacuformed fuselages, it is usually prudent to attach some interior components to the side (and lower) elements, and to paint the interior, prior to assembly of the fuselage box.

Right: The halves have been permanently joined and surface detail added to the fuselage. The basic structure has been sprayed overall with OEM grey primer, most of which has been sanded off. The engine, cowling and tail movable surfaces are being test fitted. The large cutout behind the firewall is for later installation of the tensioned monofilament nylon rigging wires: having been passed through the holes in the wings, they will pass through holes in the fuselage centerline panel and be set in place with crimped ferrules. At this point, the entire model will have been finish painted, decal markings added, oversprayed with clear lacquer and rubbed down. Thus, attachment and painting of the cover panels will be a tedious chore. (Alcorn photo)

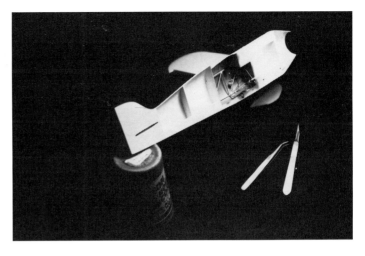

The Wedell-Williams #44s depicted herein were assembled as halves. This rather peculiar approach facilitated alignment of wings and landing gear to the fuselage. The Edison Gold Moulded cylinder record container betrays another passion, collecting vintage phonographs/cylinders and gramophones/ 78 rpm disc records. (Alcorn photo)

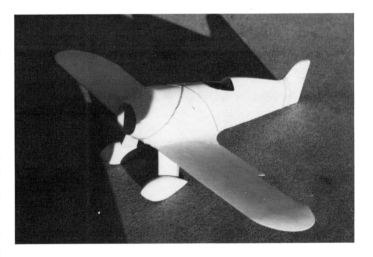

Here the halves are temporarily assembled for test fitting. The sun is about to set, into Mar Pacifica. (Alcorn photo)

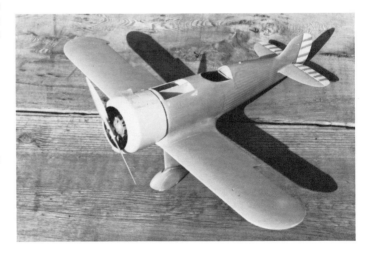

54 SCRATCH BUILT!

AT RIGHT:
Ninak Fuselage Components-1

These elements comprise the heart of the basic DH9A fuselage structure. The main purpose of this sequence is to reveal the modular, pre-fabricated approach for slab-sided aircraft. Note that the visible interior is painted, and will be fully fitted out prior to permanent assembly of the main box structure. (Alcorn photo)

Ninak Fuselage Components-2
Here the left side panel has been temporarily attached to the bottom one, a process repeated countless times during fitup. Three solid (non-visible after completion) rectangular bulkheads have been permanently attached to the bottom panel. Two 0.125 inch diameter holes in the side panels match the slightly projecting ends of the brass tubes which are permanently and rigidly installed in the bottom panel: they will later serve as sockets for the brass spars of the lower wings. Thus, during temporary and permanent fuselage assembly the side panels are snugly attached and precisely registered with the bottom panel, and hence, with each other. These fuselage tubes have been partially cut at their midpoints and bent for correct dihedral. Thus, installation of the lower wing will be "automatic" as regards location, angle of attack (3 degrees) and dihedral (3 degrees). A piece of drafting tape holds the elements together at the rear, while notches at the bulkhead edges mate with the longitudinal stringers of the side panels (which are full height, while the bottom panel is undersize by the thickness of the 0.020 inch thick side panels). (Alcorn photo)

Ninak Fuselage Components-3
This shows the exterior surfaces of the two sides, semi-finished, with simulated fabric panel and lacing grommets (tiny wire stubs), access steps and moisture distortion of their "plywood" portion. (Alcorn photo)

For many aircraft such as the Boeing 40B-4 and 80A-1, the slab sides are relieved by longitudinal stringers over which fabric is stretched. Bob Rice simulated this effect by gluing plastic strips to the box shell at the stringer locations: for the 40B-4, he used 3/64" Plastruct angle, aluminum templates providing correct interstringer spacing and alignment. Using these templates as a guide, he then embossed the stringers on 10 thou "skin." He attached the skin to the box/stringer unit with five-minute epoxy: he used Superglue on the 80A-1. Another choice for this purpose would be casein glue (Elmers).

Cabin window construction for the 80A-1 was performed in the following manner: Window openings were cut out of the 40 thou fuselage sides before box structure assembly (however, he did not cut out the large door opening until after box assembly, to avoid assymetric distortion problems). After application of the exterior skin, it was removed at the windows, using the 40 thou cutouts as a guide for the knife. The thin inside and outside window frames were photoetched brass. The frame openings were appropriately undersize relative to the fuselage cutouts, while the windows were cut from 20 thou clear plastic to just fit the cutouts. The interior frames were attached with Superglue, using a pin "applicator" from the outside, through the window openings. (In this case, the interior frames were backed with 20 thou plastic, to simulate window moldings in the actual aircraft.) The windows were then Superglued in place from the outside. Finally, the outer frames were attached, their bottom edges having been aligned to a temporarily affixed straightedge. Bob reports that both the technique and results were very satisfying.

The October 1986 *Fine Scale Modeler* has a good article on Rice's technique for the 40B-4, while his 80A-1 is well covered in the September 1987 *Scale Modeler*.

WINGS

There are at least three basic methods of wing construction available to the static scale scratch-builder: solid; vacuformed shell over internal structure; and the rib, spar and embossed skin techniques.

Chapter IV: The Basic Model Structure

SOLID: For small models, or aircraft with very thin wings, especially those having underwing camber, it is not unreasonable to carve the wings from solid polystyrene stock – after all, we use this method for the tail surfaces of even large models. The only concern is possible long term droop of long, self-supporting (cantilevered) wings.

A variation is to build up a solid wing by gluing sheet stock together, with embedded brass tubing for rigidity. But, this too has its pitfalls: I (Alcorn) laminated the upper wing of my 1/32nd scale Rumpler CIV in this manner, using MEK. Unfortunately, within a year after completion, the wing acquired a pronounced droop, despite the presence of two brass tube spars. Doubtless, residual stresses within the laminate, in concert with the underwing camber, possibly aggravated by rigging tension, was responsible for the problem. One solution, which I intend to use for eventual restoration of the Rumpler, is to carve the wing from basswood.

Whether from plastic or wood, for vintage aircraft one must simulate the fabric over rib surface contours. While fine thread can be used, our preferred method for smaller models is to glue thin strips of paper down at each rib location, and contour with an appropriate filler, sanded into shape. Cut 20 thou wide strips (eyeballed) from clear vellum, using a sharp No. 11 surgeon's (or Xacto) blade and metal straight edge. Spray the wing with OEM grey auto primer. Carefully mark the rib locations with pencil. Then, place a vellum strip over each

This 1/16th scale model by John Alcorn represents #44 as fitted with the 800 hp Pratt and Whitney WASP "C", in which configuration it won the Shell Speed Dashes, setting a new landplane world record of 305 mph over a 3 km course at Chicago on 4 September 1933, piloted by Jimmy Wedell. The end came for #44 and for Doug Davis during the 1934 Thompson, when it stalled and crashed as Davis was attempting to recircle a cut pylon. This model is a twin of the one made for the NASM, except for the engine and cowl. (Francie Alcorn photo)

This 1/16th scale model built by John Alcorn for the NASM represents the legendary Wedell-Williams #44 racer when it won the 1933 Thompson Trophy, piloted by its creator, Jimmy Wedell. P & W Wasp Jr. engine. (NASM photo)

SCRATCH BUILT!

During attachment of the skin to the DH9A wing structure, I had destroyed or distorted the rib embossment along the leading and trailing edges. (I still have a lot to learn from George.) To restore and enhance the rib (and false rib) simulation, I added paper strip atop all of the embossments. Then, I buried these in OEM grey primer, as described in the text. Finally, I generated the interrib contours using rolled sandpaper (see next figure). (Francie Alcorn photo)

OEM grey auto primer having been brush and spray applied to bury the paper strip enhanced rib embossment, a roll of sandpaper is used to create the appropriate interrib "catenary." (Francie Alcorn photo)

rib location, holding it taut by the ends, which are folded around and pressed against the wing underside. Apply lacquer thinner to the vellum strip and press it down so that it adheres well to the painted surface. The tricky part is to get it on absolutely straight. For larger models, 5 thou plastic strip, commercially available, can be used – glued directly to the plastic skin.

Once the rib strips are in place, begin the filleting process by brush painting (slopping) unthinned filler (OEM grey auto primer is perfect) over each strip and quickly squeegeeing it into place with the forefinger, run down the length of the rib. The filler is gradually built up by repeated application, with periodic sanding. The interrib fillet contours are developed by very deft fore and aft strokes with a piece of rolled sandpaper (No. 220, later 320 grit). As the proper contours begin to emerge, "finger painting" gives way to airbrush application of primer, thinned only as necessary. Eventually, your patience and skill will be rewarded by a very satisfying rib effect, with no evidence whatever of the paper foundation. Incidentally, this rolled sandpaper contouring must be done under a point light source, with the wing surface held at a shallow angle to produce a hard ridge shadow. We older guys are wearing Optivisors.

For solid plastic wings with underside camber, a reverse effect is required, with the surface slightly depressed at each rib location. To achieve this, first mark each rib location and scribe (with an appropriate tool, as for scribing panel lines). Then, using a No. 11 blade, scrape away the surface, so that it dips in a gentle slope to the base of the scribe line. This is a rather tedious process, which must be done very carefully to achieve an appearance which is crisp, delicate, and even from cell to cell.

Another good approach for solid wings is to cover a basswood core with 10 thou styrene sheet, whose rib loca-

tions are embossed, as described below under "Ribs, Spars and Skin." The skin is then attached to the appropriately undersize core, using epoxy or contact cement. Bob Rice employed this technique for both his Boeing 40B-4 and IL'YA Muromets. He now favors the contact cement (IL'YA) approach over epoxy. First, spray both skin and core mating surfaces with contact cement. When both are superficially dry, lay the skin down over the core with brown wrapping paper in between –except for a thin contact line at the leading edge. Gradually withdraw the brown paper, pressing the surfaces together as you go. Finally, leading and trailing edges of upper and lower skin are glued together, trailing edges having been feathered appropriately by block sanding.

However, perhaps the best solution for solid wings is to cast them from epoxy resin, using Peter Cooke's technique as described in Chapter V. In this case, the above comments regarding rib simulation apply to the wood master form, rather than to the final wing itself.

SHELL OVER STRUCTURE: For modelling thicker, metal skinned wings such as the 1/32nd scale A-20, vacuforming over carved wood molds is a good technique.

Making The Forms: Since both the upper and lower surfaces have airfoil, it is necessary to prepare forms for each. The procedure is generally similar to that for the fuselage. Select hardwood plank, hopefully of thickness appropriate for the upper and lower surfaces (if the source block is significantly thicker than necessary, ferheavensake, have it planed to size, or do it yourself if you have the tools). Then, make the planform template, carve or bandsaw to rough shape, file and block-sand the leading and trailing surfaces and lightly glue the upper and lower planks together. Once glued, give a final sand-block truing to the edges, to ensure that the final products match exactly.

Chapter IV: The Basic Model Structure

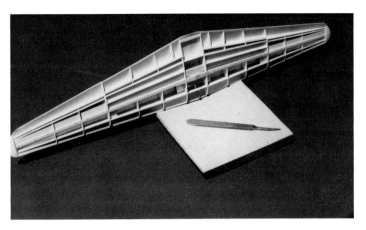

A-20A Wing Structure: Attachment of the upper skin was easy, since it was accessible from below for gluing. Some structure in the rear was added thereafter. The ailerons were made integral and later removed. (Alcorn photo)

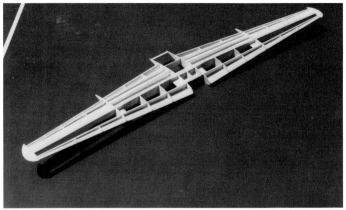

Although in retrospect probably unnecessary, the main spar was basswood, to ensure rigidity. Tips were shaped from solid (laminated) polystyrene sheet. Development of the outer section "washout" (from +3 to -1 degrees) required premeditation and careful assembly. (Alcorn photo)

Wing thickness as a function of span (front view taper) must next be generated. Large, relatively flat, wide chord surfaces are difficult to carve: a small plane can be a handy tool for this job.

Now, using the airfoil templates as guides, carve, file, and block-sand the wings to their proper cross section shape. The outer wing panels of the A-20 presented a special problem, since the airfoil featured "washout"; from +3 degrees at the root, to -1 degree at the tip.

The wing forms must of course be appropriately undersize, to account for final thickness of the shells. Twenty-five to thirty mils is the proper allowance for 40 mil sheet stock: some slight thinning occurs during vacuforming – around 5 mils is typically removed later during block sanding of the surfaces (see below).

Also, the planform (chord) size of the wing should be undersize, to allow for runout of the shell beyond the trailing edge, as well as for leading edge thickness. Trailing edge runout is a function of both shell final thickness and of the terminal angle of incidence of the upper (or lower) airfoil section. Be somewhat generous here, since you don't later want to run out of chord during final planform truing. So, use the initial sheet thickness (say 40 mil) for your calculations. Conversely, since it's being pulled straight down during vacuforming and since considerable final edge block sanding will occur, allow about half of the sheet thickness for the leading edge.

Internal Structure: For making the polystyrene wing structure, several important factors must be considered:
• Proper dihedral, rigidity and dimensional stability are achieved by one or more main spars at appropriate locations, typically those of the actual aircraft. Main spars should be fashioned from thick stock, say 1/4 inch.

RUMPLER C.IV DETAIL: The 240 hp Mercedes DIVa, in conjunction with its excellent design, endowed the CIV with outstanding high altitude performance, enabling it to usually avoid interception during reconnaissance missions. The later CVII and Rubild variants could operate at up to 22,000 feet, the aircrew breathing oxygen through mouth tubes. (Joe Faust photo)

• Various wing features requiring cutout areas, penetrations and hardpoints must be accommodated. Cutouts include ailerons, flaps, and wheel wells. These require careful placement of edge "closure" pieces for fit, strength, and neat, correct simulation of the area if later visible (as for wheel wells). In the case of ailerons (and flaps) it makes sense to install edge strips to either side of the seam, so that when the flying surface is later removed, it will have its own structure. For multi-engined aircraft with nacelle wheel wells, the underside wing skin is often removed, exposing

When mass production of identical ribs is necessary, as for rib, spar and skin construction of DH9A wings, the technique shown in this and the accompanying photo is employed. Here we see the rib being cut from a 15 thou plastic blank, using the brass template, pinned into a balsa block – a basswood cutting strip is sandwiched beneath the rib blank. A #11 surgeon's blade is used for cutting, held perpendicular to the surface. Pin holes (later enlarged to spar size) were predrilled into the rib blanks. (Alcorn photo)

Here we see the complete "tooling" for rib production, with 15 thou rib blank in place between the brass template and cutting strip. Beside it is a complete set of ribs for the Ninak top wing, assembled on stub brass tube "spars." This set is block sanded, with several "shuffles", to produce fully uniform ribs. O.K., so there are only two holes in the brass template, and three spar stubs in evidence. I have no explanation. (Alcorn photo)

structure and gear attachment points: this of course must be anticipated as the wing structure is planned.

• The wing skin must be reinforced sufficiently to prevent local distortions (sagging, waviness) after completion. This requires a rather extensive internal structure, including ribs and secondary spar elements. Typically, unreinforced shell cells should have characteristic dimensions of about 1 inch on a side or less. For some models though, wing structure should show subtle evidence of ribs, etc., beneath. In such a case, ribs should be placed in correct locations, without skin bracing in between. I

(Alcorn) built the wings of my 1/16th scale Wedell-Williams #44 racer in this manner to simulate the appearance of the plywood skin undulations evident in numerous photographs of the real aircraft.

While the complete wing structure could be made independently, it is usually convenient to add some of it after attachment of the upper shell. Most of the secondary stiffeners can be "custom" shaped and fitted as you proceed, rather than attempt to predetermine the geometry by layouts.

Shell and Structure Preparation: The trailing edge of the

This view shows positioning of the premarked skin for embossing of a rib. A piece of manila paper, and a 45 degree triangle have been clamped upon a clipboard. The skin piece is placed against the hypotenuse edge of the clamped triangle. A second triangle is set for embossing the next rib location. Note the embossing tool in the foreground. (Francie Alcorn photo)

Here a rib is being embossed. With one hand, the floating triangle is held firmly down, care being taken that both it and the skin are against the edge of the clamped triangle. With the tool held perpendicular to the edge of the floating triangle, embossing is performed with one firm, smooth stroke. (Francie Alcorn photo)

Chapter IV: The Basic Model Structure

VOUGHT-SIKORSKY SB2U-2 "VINDICATOR": Another fascinating late prewar US Navy subject rendered by Arlo Schroeder, this model representing an SB2U-2 of VB-2 serving aboard the carrier Lexington. At the 1982 IPMS/USA Nationals in St. Louis, Arlo's Vindicator was voted Most Popular Model, as well as receiving the Detail and Scale award and taking first in the Scratchbuilt – 1/48 and larger category. (Photo courtesy Arlo Schroeder)

SB2U-2 DETAIL: Here is a glimpse of the attention to detail lavished by Arlo Schroeder upon this scout/dive bomber. Its only significant combat operation occurred in June 1942 when Midway based Marines attacked the retiring Japanese armada on the 4th – one stricken Vindicator crashing into the cruiser Mogami. (Photo courtesy Arlo Schroeder)

completed wing must be straight, sharp, strong, and accurate in plan view. Achieving this result requires considerable planning – and effort:

• After trimming (slightly oversize in chord), the vacuformed wing shells should be block-sanded over their supporting wood forms, in order to begin the process of truing the surface.

• Then, the shell trailing edge is bevelled on its underside with a large sanding block, while supporting the outside edge with some appropriate straight surface such as a thick block or table edge. In order to ensure that the bevelled surfaces of the upper and lower shells mate at assembly, each should be flat – that is, lie in the wing split plane. This can be achieved by lightly rubbing each shell on a large, flat piece of sandpaper from time to time. This technique shouldn't be used for primary bevelling though, since it's tough to retain uniformity of material removal along the wing. A trick way of trimming the shells to the split plane is with their wood forms. Simply separate the form from its base, insert it into the roughly trimmed shell and carefully block-sand until leading and trailing edges of the shell are flush with the base of the form.

• The completed wing internal structure (ribs, etc.) must "feather out" at the forward edge of the top shell trailing edge bevel. Also, the overall structure must be smoothly contoured so that it will mate to the lower shell without gaps or humps. This is achieved by block-sanding the structure underside in spanwise strokes, taking care not to damage the trailing edge bevel or leading edge of the top shell.

Upper Shell Installation: This step is comparatively straightforward, since the shell to structure interfaces are accessible from beneath for gluing. Nevertheless, the shell should be test-fit over the structure beforehand to ensure good contact at all appropriate locations. Then, after the shell has been attached, but before the glue joints have dried, carefully check for any distortion. To assure straightness of the trailing edge, it should be held firmly down upon a flat surface until the glue has thoroughly dried.

Lower Shell Installation: We are now at the point of installing the underside shell of the wing. This is rather challenging for large wings, since the mating elements aren't accessible for MEK brush/capillary application except along the leading and trailing edges.

The first step is to carefully test fit the lower shell to the structure/upper shell. A fair amount of such fitting and minor trimming may be required to assure yourself that all elements are in contact when the lower shell is pressed into place. For larger models such as the 1/32nd scale A-20, the wing panels are in two pieces: inboard and outboard. This facilitates the process considerably.

Plastic cement from a tube should be quickly and liberally applied over the mating edges of the structure and the lower panel pressed into place – it must, of course, be held there until the glue has set. Then, fuse the leading and trailing edges together with MEK. Sight down the trailing edge to see if it is straight. If some minor waviness is evident, press it down firmly against some straight surface. Now, despite your best efforts, the lower panel may not be firmly attached to the structure over some areas, as evidenced by "squishiness." All

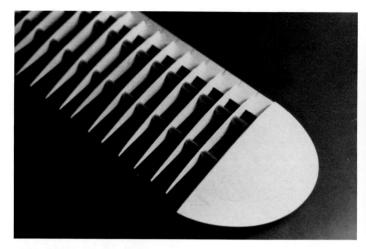

This is the unskinned structure for my (Alcorn) 1/16th scale Laird Super Solution. Holes have been drilled into the preformed tip for insertion on the brass tube spars. The ribs have been located by and attached to the premarked and shaped leading edge strip. (Alcorn photo)

The upper skin, with pencil-marked rib locations, has been installed and the preshaped trailing edge piece then applied. Note the false rib for later attachment of the interplane strut. (Alcorn photo)

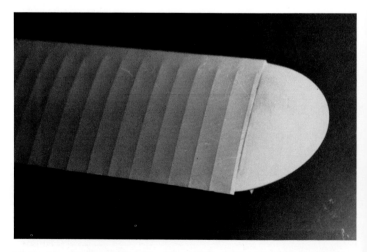

This view shows rib embossing of the attached wing upper skin. (Alcorn photo)

Chapter IV: The Basic Model Structure

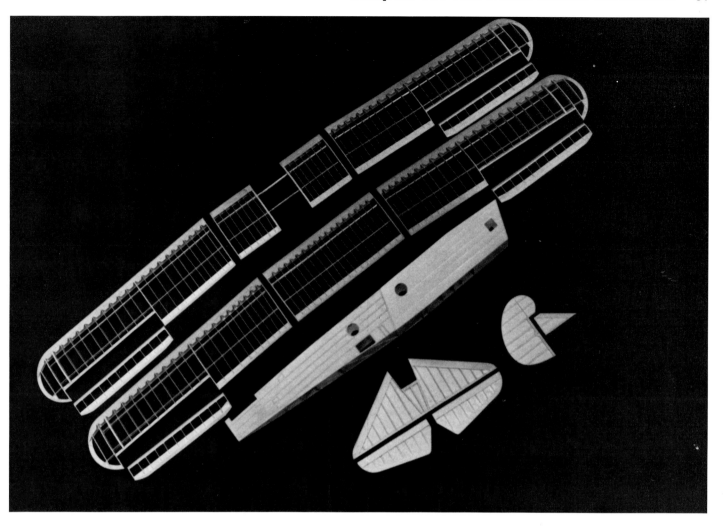

This elegant composition shows the primary structural elements of George Lee's Keystone B4A. This shot offers incontrovertible proof that his *magnum opus* was "in the works" for at least 15 years, since it is on the same film strip as photos of my Douglas A-20A wing structure, all taken in 1973. To be sure, he completed other modelling projects between then and the 1988 Nationals. (Alcorn photo)

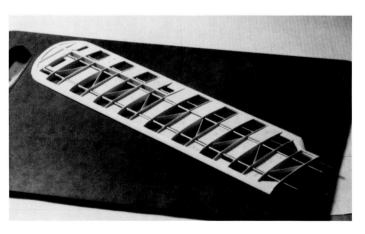

Here is the completed DH9A wing structure. For this project, extensive use was made of interrib shear panels and braces for precise rib location, as well as rigidity. I had experienced distortion problems with a prior attempt at Ninak wings, so perhaps overreacted. (Francie Alcorn photo)

you can do is drill a series of small holes through the shell where you know that structure exists beneath, such as the spar. Then, dab MEK into each hole until things stick together, and later fill the holes.

Final Shaping: While the wing is now structurally complete, some fairly serious final truing of the upper and lower surfaces may be required. First though, check the planform, which should be slightly oversize at the trailing edge and will probably need some minor truing along the front. Carefully block-sand these edges until the final shape is achieved. Now, with a large, flat sanding block, coated with rough sandpaper (180 or even 120 grit), work both surfaces using spanwise strokes. Airfoil templates can be used for checking, although at this point only minor corrections can be made. The main thing now is to get the surfaces straight spanwise, smoothly contoured chordwise, and the trailing edge straight, even and thin. Naturally, progress to finer grades of sandpaper as these goals are approached.

Removal Of Flying Surfaces: Now carefully mark the aileron and flap locations. Using a small metal straight edge

SCRATCH BUILT!

The Napier Lion engined Victoria was a troop carrier adaptation of the Virginia bomber; 96 examples being constructed between 1922 and 1933. Enjoying long overseas service, the Victoria's greatest claim to fame was participation in the first major airlift, when eight No 70 Squadron aircraft helped evacuate 586 civilians and their baggage from Kabul during the 1928-1929 Afghan crisis. This impressive 1/36th scale model was constructed by Alan Clark. (Nicholas Nikiforakis photo)

(such as a 6" pocket rule) cut away these elements with a #11 blade. Cleanup and shape the edges with small sandpaper blocks – some minor repairs may even be in order. Further, one of the mating surfaces may require the grafting of a plastic strip in order that, when reinstalled, the flying surface is flush with the wing trailing edge.

RIB, SPAR, AND SKIN CONSTRUCTION: This is the George Lee technique for simulating fabric covered wings. Here, the wing structure closely follows that of the real thing. Basically, the structure consists of plastic ribs mounted onto (usually) two brass tubing spars: usually it is appropriate to also include leading and/or trailing edge strips and tip pieces. The skin can be either flat or vacuformed polystyrene sheet, with the rib locations embossed on the inside surface.

Ribs: Let's consider the usual case, in which all, or most of the ribs are identical. A template is required which can be used for cutting out all of the ribs from polystyrene stock (typically 20 mil). The best template material is 10 mil brass sheet: it's tough to cut out with a knife, so we prefer to rough it out with snips and file down to exact shape. The template should be slightly smaller than the full airfoil shape, to account for thickness of the skin (shell). However, some allowance must be made for collective block-sanding of the cut ribs (see below). So, make it about 5 mil undersize all around. Also, the chord should be undersize to allow for the trailing edge seam joining the upper and lower shells. Select spar locations and drill a small hole in the template at each.

Cut out the requisite number of rib blanks (plus a few extras) from 15 or 20 mil styrene sheet. Drill two holes in each rib blank, using the brass template as a guide. Then, one by one, pin the template down over the rib blank, which is backed by a basswood strip to serve as a "cutting board." This assemblage is mounted upon a thick balsa block which serves as a base and receptacle for the straight pins (see photo). The same two vertical "post-holes" in the balsa block should be used for all ribs.

Holding the template firmly down against the blank, cut out the rib with a No. 11 blade held vertically (watch that

Chapter IV: The Basic Model Structure 63

NAVY-WRIGHT NW-1: Two NW-1s were constructed as test beds for the 650hp Packard V-12 T2 engine. The participation of "9" (A 6543) in the 1922 Pulitzer race was less than auspicious. On the fourth lap, USMC Lt. Sandy Sanderson went down in the shallows of Lake St. Clair after his engine failed, barely extricating himself from the overturned aircraft. (Sanderson went on to achieve legendary fame, partly as CO of VF-9M). The NW-1s streamlining and engine power was severly compromised by its lower wing, Lamblin radiators and profusion of struts. This lovely model is the handiwork of Alan Clark. (Nicholas Nikiforakis photo)

---

clamping finger!). After all of the ribs are cut out, drill out the spar holes using successively larger drills – the pin holes serve as the "center punch."

Feed all of the ribs onto one brass spar (or shorter tube section of the same diameter). Pack them tightly together and ream the other hole with the full size drill. Now, insert the second brass tube section. You now have a "solid" (actually, laminated) wing segment of length equal to the total rib thickness. It is probably damn tight, so it should hold together for the next step.

Block sand the rib pack: if you've cut them properly oversize, all should be sanded before you reach the final rib size and shape. The ribs should be removed from the brass tubes (two at a time) and shuffled two or three times during the block sanding process, to ensure uniformity. Now they're all exactly alike – you have true production interchangeability, as first achieved by Cadillac in 1909!

Leading and Trailing Edges: A leading edge strip is always required. Trailing edge strips may not be added to certain thin wings such as those with scalloped edges (wire stretched between ribs on the actual aircraft).

Such strips are made from thick (40, even 60 mil) styrene strip, sometimes laminated to form "stepped" back edges. While rather a tedious process, both leading and trailing edge strips should be fully contoured in cross-section before wing assembly. This is best achieved by block sanding the strip, which is supported along its length by a table edge (or equivalent). For bulkier leading edge strips, this process can be preceded by careful "whittling" with a No. 11 blade.

Skin: If the full upper surface contour is desired, including wing tips and leading edge, then a shell should be vacuformed over a basswood form. Typically, the lower surface can simply be made from unformed sheet stock. If the choice is to make separate tips and use a leading edge strip, then the upper surface can also be made from unformed sheet, 15 mils thick for larger wings, 10 mils for smaller ones.

SCRATCH BUILT!

Accurately mark the rib locations on the skin underside – this is more difficult of course if the shell has been vacuformed. Purpose made embossing tools are available – they have a little steel ball on the end – but a ball point pen can be used. Set the skin on a thin pad of 8-1/2" x 11" paper atop a hard, flat surface. Using a pair of drafting triangles, emboss each rib location as indicated in the accompanying photos. Practice will be required in order to produce rib grooves of appropriate and consistent depth.

Structure Assembly: We have found that it is best to assemble the wing structure directly on a print of the wing drawing.

First, feed the ribs onto the two brass tube spars cut to the appropriate length. The ribs at the aileron locations must be cropped to receive trailing edge strips. After cropping, the ribs should be assembled upon two short brass tubes for block sanding of their trailing edges, to ensure perfect uniformity. Then, feed them out the actual wing spars.

Very accurately mark the rib locations on the leading edge strips. Then, place the leading edge on the drawing and tape it down along its leading edge, so that during the assembly process, the wing structure can be rotated up ("hinged") for certain local gluing, and then laid back down with perfect registry.

In the standard George Lee approach (see photo of his Keystone wing structure), the ribs are constrained longitudinally only by attachment to the leading and trailing edge strips: as the upper skin is applied, each rib can be moved slightly with tweezers until it slips into its embossment groove. I (Alcorn) found that this worked splendidly for my robust Laird Super Solution wings (see photos). Use of interrib shear panels is a "belt and suspenders" option for stiffening the structure and locking the ribs into position before skin attachment. In this case, the skin embossments must exactly match the rib locations beforehand: a non-trivial "quality control" operation. I used shear panels for the thin-ribbed wings of my Ninak (see photo).

Prepare interrib "shear" panels (20 thou stock) whose width is cut precisely to the rib spacing less one rib thickness. As shown in the accompanying photo of the NINAK wing, interrib shear panels should be installed at every other bay, to later allow access to every rib for gluing on the upper skin. Note that three shear panel sections will be installed at each strengthened bay: leading edge, trailing edge, and interspar.

After the wing trailing edge strip at the aileron location has been installed, lightly glue in place a spacer strip to provide the proper wing to aileron gap when later removed. To it then lightly attach a third strip, which will become the aileron leading edge. Then, carefully shorten the rib rear portions which were previously cropped. The aileron structure is completed by installing these between the leading edge strip and the wing trailing edge, which has not been severed from the wing inboard region.

For the tip structure of the NINAK, I added a 30 thou shaped piece, notched to receive the spar ends, as shown in the photo. It is MEK'd to the leading and trailing edge strips, as well as to the last true rib. "Fake" rib pieces are added to support the embossed top skin at the outermost rib location.

The completely assembled wing structure should be lightly and carefully block-sanded to ensure that the ribs are all flush with the leading/trailing edge strips. Some time during all this, add hard points for anchoring the interplane/cabane struts, etc.

The skin can be faired into the leading edge strip by providing a longitudinal step in the strip, or by overlapping the skin and later feathering it to the leading edge by block sanding.

Upper Skin Installation: The upper skin is first glued to the wing root rib. Very careful fitting is required at this point to assure that the skin embossments are matched exactly with the rib/leading edge locations – and that the fore to aft alignment is perfect. Once this has been achieved, the skin can be brought down against the structure, and MEK'd in place from the underside, rib-by-rib.

Sounds simple, but there are pitfalls other than misalignment. One is warping of the wing due to stresses induced by the skin and effect of the MEK on the leading/trailing edge strips. So, wing straightness must be checked during skin application, and the wing taped down to a flat surface as it dries. *Praemonitus praemunitus.*

Lower Skin Installation: Installation of the wing underside skin presents a special challenge, particularly if the airfoil is cambered. The main difficulty of course is that the ribs are inaccessible for gluing, except by peeling back the skin, with the attendant risk of cracking or distortion.

The dope-shrunk linen on the underside of cambered

VOUGHT O3U-3 DETAILS: Not content to rest upon his Keystone laurels for very long, George carried the Corsair concept a step further, by faithfully replicating the fuselage structure with soldered brass tubing, and providing cockpit actuation of the movable rudder, elevators and ailerons. Next step, a working scale engine? (Alas not to be, since George passed away shortly before these words were penned. At his request, I will complete the model – J.S.A.)

Chapter IV: The Basic Model Structure

Here, thread has been applied to the vacuformed shell of the 1/16th scale Wedell-Williams #44 fuselage. In this case, the stringer locations were marked on the bare plastic with pencil and the thread "stuck" in place with brush applied MEK. Brush/finger application of unthinned OEM grey auto primer to develop the interstringer contours has just begun. When enough primer has been laid on, the "wave" contours will be developed using tightly rolled sandpaper. Perceptive viewers may have detected Nipper, Francis Barraud's faithful terrier, who achieved posthumous fame listening to "His Master's Voice" issuing from a Berliner "Improved Gramophone." (Alcorn photo)

wings bulges out between the ribs. This effect is best simulated by embossing the 20 thou skin on its exposed (underside) surface. Begin attachment at the inboard rib, alignment precautions being taken as for the upper skin. Progress outward one rib at a time, allowing 15 minutes or more between each for initial drying. Gently but firmly press against the prior rib with the thumb of your "holding" hand. Pull back the skin just enough to allow brush access to the next rib. The thoughtful reader will note that three hands are thus required for this operation. Should you be blessed but with two, the skin can be held back by a table edge, or some such projection. (You could, of course, enlist the services of mom, a sibling, or spouse. However, moms tend to be fluttery; siblings are notoriously unreliable; and the spouse is liable to become amorous, impatient, or both.)

For brushing the rib underside edges, Testors (or equivalent) Plastic Cement is recommended. While it contains MEK, it is less volatile, drying more slowly. Apply it copiously along the rib edge, taking care neither touch the skin, or let a drop fall onto the skin elsewhere in the interrib cell.

SURFACE FEATURES

No matter how well the basic structure has been formed, its surface must convey the appropriate appearance, be it fabric, plywood or metal. Kits have improved dramatically over the years, from rivets which would scale to those seen on the Brooklyn Bridge and panel lines which you'd either trip over or fall into, to the crisp and relatively convincing tracery seen on a modern quality offering. (We can all have a good snigger at the raised insignia on certain old kits that you could irrigate with the appropriate colors.) But, scratch building provides the opportunity to carry surface effects beyond the levels of realism achieved by any kits so far.

FABRIC: In the Wings section above, we discussed methods for rib simulation of fabric-covered wings. But that's simply one aspect of replicating the appearance of fabric covering. After all, most early aircraft fuselages employed fabric stretched over a wood (or metal) frame, at least for the rear portion. One need only consider the Hawker Hurricane to realize how long this practice remained in vogue. It still comes as a shock to recall that the outer wing panels of all WWII era F4U Corsairs were fabric covered aft of the main spar. But, the ultimate in fabric covered anachronisms was the redoubtable Vickers Wellington, whose Barnes Wallis-inspired geodetic construction could be discerned beneath its linen epidermis.

While not as dramatic an effect as fabric-covered wings, the dope tautened fabric of slab-sided fuselages revealed the underlying structure with which it came into contact. Thin strips of paper can be laid on and "feathered" in with filler to simulate interlongeron struts. Also, thin paper with well defined edges can be used on larger-scale models to indicate fabric reinforcement.

Now, *regardez vous* almost any photo of an operational WWI type having a slab-sided, fabric-covered fuselage: the SE5A, RE8, BE2 series, FE2B, and DH2 are archetypal. While much of the surface may have retained its drum-like tightness, wrinkling is almost inevitable in certain regions, especially where load induced distortion has occurred. While such wrinkles take many forms, a coherent pattern is usually in evidence, where membrane tension is locally relieved in one direction but not in the other – as opposed to the complex folding of loose clothing or drapery.

While simulation of fabric wrinkling may at first seem daunting, it can actually be fun, once you get the hang (stretch) of it. As a prelude, it is wise to locate a photograph of the aircraft type in question, displaying the fabric pattern which you intend to replicate. First, simply sketch the pattern on the plastic surface with a pencil: it can, of course, be erased many times before the desired pattern is obtained. Then, using a curved edge knife blade, begin cautiously scraping away – holding the fuselage so that the incident light casts distinct shadows. Once the convolutions have been roughly generated, sandpaper is used to achieve the final, smoothly transitional effect. The process can be quite artistically satisfying – and will add greatly to the realistic look of the finished model. Beware of a pattern which is unnaturally rigid and even – or too pronounced. As with painted weathering effects, it can all too easily be overdone.

On most aircraft with fabric-covered fuselages, certain panels were detachable – or at least capable of being folded back for access to the interior. Lacing was the usual means of attachment, the cords passing through grommets or over hooks. I (Alcorn) faced this challenge on my Ninak, whose removable panel covered three rear fuselage bays: for this type, hooks were used throughout. I very carefully scribed seam lines and laid out the hole pattern to receive the "hooks" on the side panels, prior to their assembly as the fuselage shell. This made layout and hold drilling easier – besides, if I really messed up a panel, it would be a lesser disaster. The hook holes were made with a small shank drill, mounted in a pin vise. Since I could conceive of no way to uniformly produce 1/24th scale cord hooks, I settled for small diameter brass wire, around which I could later pass fine nylon thread. Thin paper was used to simulate the cloth reinforcement strips on either side of the seam. Tedious, but the end result was reasonably convincing. For stringer simulation on double-curved surfaces, the thread and filler method is your best bet; whether as a master form for resin casting, for a solid wood model or upon a plastic shell. (Upon first thought, it may seem logical to produce the stringer effect on the forms over which shells are to be vacuformed. However, it is not possible to achieve sufficient sharpness of line through 20-40 thou stock.) In any case, proceed as follows:

The surface is first given a good coating of grey auto primer. The stringer locations are then sketched on with a pencil: you'll doubtless erase repeatedly until the correct spacing and "flow" is obtained. Still taking the Hurricane as our example, the turtle decking behind the cockpit is the most challenging. Anchor each thread at its forward end and work aft, laying the lightly tensioned thread down over the pencil line, and gently mashing it into the primer, which has been softened by brush-applied lacquer thinner. As the threads converge at the rear, frequent adjustment must be made before adequate spacing and contour is achieved, despite your best pencil work.

Once they're all in place to your satisfaction, begin laying on thick OEM grey primer using the brush and finger method, as described above for wing interrib buildup. Eventually, you'll arrive at the point where sanding is in order to begin truing the interstringer contours – ultimately to result in a series of "pointed waves" in cross-section. This is achieved by using (initially) 180 or 220 grit sandpaper, rolled into a cylinder of the appropriate diameter and held with the thumb and forefinger – you'll need progressively tighter rolls as you work aft, although the depth of the contour should steadily diminish. While this procedure is tedious and trying of your patience, a beautifully sculptured pattern will begin to emerge. Additional primer must inevitably be added here and there, applied by brush and smeared with the forefinger as before. Towards the end, you'll need to add several airbrush coats of primer – partly to obtain saturation/coverage of the threads. No vacuformed shell and precious few kits can achieve what you can by this process.

PLYWOOD: In principle, replication of plywood-skinned surfaces should be easy, since the underlying structure is obscured, rivets aren't present and panel lines are rare or nonexistent – thanks to the technique itself, use of filler, and often, fabric covering. In practice however, this smooth, relatively featureless effect is achieved only by factory fresh examples and by such latter-day "wooden wonders" as the Lockheed Vega/Orion series, De Havilland Mosquito and Hughes "Spruce Goose."

But for operational WWI era machines, it's quite a different story. German aircraft especially made extensive use of plywood covering; for lightness and strength, if not for ease of manufacture. Exemplars which come to mind are the Albatros D.I – D.Va and C.V-C.XII series, Pfalz D.III, DFW C.V, LVG C.V/C.VI, Roland "Walfisch," Hannover and Halberstadt CLs. While some, including the Pfalz D.III seemed to retain their factory fresh appearance quite well, most soon showed evidence of panel distortion in varying degrees, due primarily from moisture absorption. Classic in this respect was the slab-sided Albatros D.III, though equally exotic patterns can be discerned on photographs of Halberstadt CLIIs and IVs, and DFW CVs.

In the case of plywood, we're observing various manifestations of buckling phenomena, as the once taut panel loosened from moisture-induced expansion. Where it was attached to the underlying structure, the correct form was preserved, but elsewhere it expanded: being unable to grow in the plane of its surface, it deformed transversely, in or out. Such patterns are subtler than those of fabric wrinkling, requiring even more attention to photographic evidence to convincingly replicate. Again, the pattern can be sketched upon the plastic surface and realized through judicious scraping, combined with painted on applications of thick primer or "putty" for the high spots. The final effect, of course, requires careful sanding, interspersed with airbrushed primer.

METAL SURFACES: We congenital vacuformers readily defer to Peter Cooke's resin cast technique for realistic simulation of metal skinned surfaces; particularly of the WWII era, where panel lines, rivetting, Dzus fittings and lap joints were strongly in evidence. Yet, his 10 thou skinning of master forms for casting can certainly be applied to the vacuformed shells of one-off models. For that approach, we refer you to Chapter V. Even solid wood models can be given a skin of 10 thou panels, heat formed over the very surfaces upon which they will later reside – glued in place with Elmers. Aside from producing a state-of-the-art surface appearance, this hybrid technique would eliminate the age-old scourge of solids – eventual mini-cracking as the wood ages.

Nevertheless, there have been many vacuformed models sporting convincing metal surface simulation, achieved in the following manner. Panel lines are carefully laid out upon the bare plastic surface, followed by incising with a No. 11 blade and chasing the cut with a conical-ended scribing tool: the latter produces a "V" groove – the depth sets the width. This apparently simple process is complicated by the fact that many panel lines, such as those following cross-sections of a double-curved fuselage, cannot be incised using a flexible straight edge as a guide. For this, the best approach is to fabricate stiff plastic templates, backed with extension pieces which, when laid against the vacuformed shell surface, cause the template to stand perpendicular to the fuselage axis.

CHAPTER V: RESIN CASTING

PREFACE

Resin casting is a technique which is finding increasing favor among modellers for producing smaller, complex parts, especially where multiple identical elements are required – such as cylinders of radial engines. Some modellers have extended this technique to include entire fuselage, nacelle, and wing shells. Peter Cooke of England has advanced the state-of-the-art of resin casting and simulation of metal skin surface effects to new heights, as evidenced by photos of his models herein. We are indebted to him for the contents of this chapter. While the rest of us whose work is featured in this book have occasionally produced models on commission for museums and individuals, Peter is a professional modeller, producing "runs" of 5 or 6 of a given type – although no two are exactly alike as regards finish, markings, details, or even mark. For example, his Spitfire series ran the gamut from MKIX through Griffon engined MK XIX.

INTRODUCTION

I have chosen 1/24th as the scale for all my scratch-built models, since I consider this to be the smallest in which I can represent all the details visible on a full-size machine. As an engineer, with an interest in how an aircraft works, and is put together, I also like to depict internal details where these can be made naturally visible. One of the fascinations of scale aircraft building is the different directions from which people approach the hobby, and I appreciate that much of what I describe will be of interest to only a small proportion of modellers. For example, it would be inappropriate, not to say impossible to represent the lap-jointing of fuselage panels in any scale much smaller than 1/24th. I have made no compromises in what I describe, however, since I am sure that readers are quite capable of selecting what is appropriate to their own scale and level of skill. I would plead though that you do not underestimate what you can do. I blush when I look at models that I built only a few years ago, and which I considered quite an achievement at the time.

I cannot stress too much the need to examine closely real aircraft wherever possible. It is all too easy to develop in the mind an idealized image of what an aircraft should look like, based entirely on study of other models. I have seen several models which are beautiful works of art, but, particularly when photographed, look nothing like real aircraft. These errors lie in such subtleties as the surface finish, the quality of fit and finish of panels, and the feeling of weight conveyed by the "sit" of the undercarriage and the compression of the tyres. I firmly believe in photography as the ultimate test of a model. When examining the actual model, one automatically allows for compromises in view of its small size. A photograph, however, could be of the full-size machine and so the slightest error in texture of paintwork or out-of-scale details, and the eye will latch on to it straight away.

I do not pretend that my methods are "definitive": they are simply the ways that I have solved problems as they have arisen. Similarly, I can only write with authority about models I have actually built, but the techniques should be applicable to modelling any aircraft made since the beginning of World War II.

MATERIALS USED FOR RESIN CASTING

Skinning:

Plasticard: Manufactured by firms like Slaters and readily available from model shops, this consists of polystyrene sheets from 5 – 60 thou in thickness. It can be easily formed

AVRO LANCASTER: This 1/24th scale Cooke epic represents ED953, PO-Q of No. 467 Squadron, RAAF. (Cooke photo)

68 SCRATCH BUILT!

Chapter V: Resin Casting

after heating to 250-275 degrees F. It normally has a slightly rough surface which means that it has to be rubbed down and polished before it can be used to represent alloy skinning. But it is perfect for fabric.

P.V.C. (polyvinyl chloride) Sheet: This material is manufactured specifically for vacuforming. I obtain mine from a local firm that produces blister packaging. In a thickness of 10 thou, this material is ideal for producing clear cockpit canopies. In fact, I now use it for virtually all my skinning panels as well. If a mistake is made in moulding, reheating will usually send it back into a flat sheet, ready for a second try.

Moulding:

Plaster of Paris: This cavity mold material is available from any art supply store.

Silicone Rubber: This is a liquid with the consistency of cream. A catalytic hardener is added, and the mixture cures over a period of about 12 hours depending on temperature. The mould is flexible, yet stable dimensionally, and tough. It is eventually attacked by any resin, but is ideal for a batch of up to a dozen castings. Its one disadvantage is its high cost, but, if a mould is well designed, not much rubber is used.

I am currently using two forms of the rubber:

- R.T.V. 11 (white) available from Alec Tiranti Ltd., High Street, Theale, Reading, Berks, or General Electric, R.T.V. 31. This is a fairly stiff rubber when set, and therefore ideal for large castings, or ones in which there is not too much undercut.
- Dow Corning Q3-3481. This is a softer rubber, but very tough and flexible. It is therefore ideal for situations where the mould has to be stretched to remove small castings of complex shape. It has a fairly thick consistency when poured, and therefore tends to entrap air bubbles unless great care is taken.

Vinamould: This has the consistency and feel of jelly cubes and melts at 170 degrees Celsius. It is cheap and re-usable. It cannot, of course, be used on any master pattern that might melt or crack with the heat as the mould material is poured on. It is normally only suitable for making one casting since it tends to break up as the cast item is removed. Unlike with the other mould-making materials listed, it is best to cover the master-pattern with a release agent such as wax, so very fine detail cannot be produced.

Latex Rubber: This is a thin white creamy liquid that can be painted on to the master-pattern with a brush. This then cures in the air, although the process is very much speeded up with gentle heat in an oven (60 degrees Celsius is quite sufficient). Normally several layers will have to be built up to produce a strong mould which can then be peeled off the master-pattern like a glove. It is thus ideal for figures with an open base. Because of its great flexibility, a fair degree of "undercutting" can be incorporated in the master-pattern. It is not dimensionally stable enough for casting thin items like propeller blades.

Casting:

Polyester Resin: This is a strong smelling, normally clear liquid to which a small quantity of hardener is added. It then cures in about half an hour depending on temperature. It is the basic bonding material in fibre glass construction. Consistency can be thickened by mixing in a filler such as talcum powder. This not only makes the resin more economical to use, but makes a casting with slightly softer, less brittle machining characteristics. Castings made in this material are quite strong and rigid. The major disadvantages are that its quick setting does not always allow enough time to remove bubbles from the mould's surface after pouring in the resin, and also the material shrinks by about 2% during curing. This does not usually make a serious change in dimensions, but, since the shrinkage occurs before the resin is fully cured, the pulling away from the surface of a silicon rubber mould tends to mar any fine surface detail on the casting. For this reason, it is used primarily for master forms to which skinning will be attached.

Epoxy Resin: This is more expensive than polyester resin and can be more difficult to obtain. It consists of a golden liquid resin to which a normally brown liquid hardener is added. Curing then takes place over a wide variety of different times, from about 1/2 hour to several days, depending on the type of hardener and the temperature (slight warming considerably speeds up the process). This control over the setting time can be very useful if complex castings are being poured. The final casting has a slightly rubbery, less brittle nature than polyester castings, particularly in thin sections. The degree of flexibility can be increased by using less hardener. Since negligible shrinkage occurs during curing, very fine surface detail can be incorporated in the mould, and an excellent surface finish obtained.

MAJOR COMPONENTS

INTRODUCTION: My technique employs the following basic steps, which will later be described in detail:
- Preparing the pattern
- Plaster of Paris cast
- Solid polyester resin form
- Formed sheet ("skin") panels
- Skinned master form
- Master mould
- Resin cast component

Opposite:
ABOVE – SPITFIRE Mk.XIV: Where a modelling career began. Peter Cooke's first scratch-built effort; the Griffon engine/installation grafted onto an Airfix 1/24th scale Spitfire Mk.I – other mods as appropriate. His efforts were rewarded by the John Edwards Trophy at the 1976 British IPMS Nationals. (Cooke photo)

BELOW – HAWKER TEMPEST Mk.V: Peter Cooke's first completely scratch-built model, from MAP drawings by Arthur Bentley. 1977 British IPMS National Champion. On the photo back side, he comments, "I blush to look at it now." (Cooke photo)

SCRATCH BUILT!

Above & above right – HAWKER SEA FURY FBII: The model depicted in these two photos represents a Royal Navy Sea Fury, operational during the Korean War. (Cooke photos)

This general approach offers three fundamental advantages, relative to "traditional" vacuforming of shells, as described in Chapter III.

First, casting the final product in a master mould allows the creation of subtle surface features (panel lines, hatches, rivet detail, and pillowing, etc.) which are not obtainable through vacuforming, except by subsequent tedious handwork (some effects in fact simply cannot be achieved at all).

Second, "limited production runs" can be made from the master mould; and/or variations can be generated with relatively little extra effort by revising the formed sheet panels on the resin form, and casting another master mould.

Third, the final cast resin component is inherently strong and rigid, so that no internal structure is required. However, the flip side is that, if extensive interior structure is to be visible, such effects must be created by preparing cores for later removal from the final resin casting.

PREPARING THE PATTERN: Certainly the basic component shape, a fuselage half for example, can be carved from basswood, as described in Chapter III above for vacuforming. While this may be entirely appropriate for smaller 1/24th scale components, including the entire fuselage halves of single-engine fighters, I prefer the following approach for large components:

Cut out two identical fuselage side profiles from balsa sheet. Build up one fuselage half and then the other, as follows: Attach a side profile to a flat board of convenient size, using rubber cement or double sticky-sided sellotape. Pre-shaped balsa cross section (half) formers are then glued to the profile. Their size must allow for the thickness of the side profile, as well as for the balsa stringers which are then applied in a generally fore and aft direction.

The number of formers used depends on the complexity of the fuselage curves, but remember that the balsa stringers will automatically follow a smooth curve. Gaps between the stringers can now be filled with Interior Polyfilla. This has the advantage of a similar consistency to balsa when sanding, making it easy to blend the two materials together. It can be an advantage to insert a horizontal plasticard profile of the fin and rudder at a strategic point, as well as a similar vertical profile at the rear of the fin. With sandpaper wrapped around a balsa block, the entire fuselage can now be quickly smoothed to the correct shape. In the area of the fin, the scraping of the hard plasticard inserts will warn when the correct shape has

Chapter V: Resin Casting

been obtained. The entire fuselage is now painted with catalyzed polyester resin, which will soak in and set hard, enabling a really smooth surface to be obtained by final sanding. This composite element is now fully equivalent to a solid hardwood pattern.

A variation on the above is to make the side profile and cross section formers out of plasticard, balsa blocks being placed in the interformer spaces (grain being perpendicular to the side profile). The balsa is then block-sanded just to the plasticard formers, at which point Interior Polyfilla is applied to fill the inevitable balsa to plasticard gaps. Then, the pattern is sanded smoothly down until the plasticard formers are again contacted, no stringers being employed. As before, the fuselage is now painted with catalyzed polyester resin, which is given a final sanding after setting.

The fuselage halves are now temporarily glued together with rubber cement or double sticky-sided sellotape. Final light sanding is then performed across their common profile to assure a perfect match.

CASTING THE SOLID POLYESTER RESIN FORM: In principle, either the solid wood or composite patterns described above could be used for subsequent skinning. However, especially for production runs, it is advisable to cast a solid, single piece component, as follows:

Using rubber cement or double sticky-sided sellotape, reattach the fuselage half to a flat board, with a piece of "cling film" in between. The cling film must extend beyond the pattern far enough (approximately 2") so that its edges can be folded up to form a watertight "casting box." With a thin coating of wax mould release on the pattern, and its sides supported with books, the casting box can now be filled with plaster of Paris.

After the plaster has set, remove the pattern and allow the cast to thoroughly dry before coating its cavity surface with Vaseline.

Catalyzed polyester resin, powder-filled to produce a porridge-like consistency, is now poured into the inverted mould, over which a flat, release agent coated board is placed. The two solid half fuselages made in this way can then be superglued together, smoothed and polished to yield the final master form.

WING PATTERNS AND FORMS: While the process for preparing the wing patterns and casting the master forms is similar to that for fuselages, numerous detail differences require explanation.

For smaller aircraft, including 1/24th scale single engine fighters, the wing patterns can be carved from basswood, as described in Chapter III. However, the following procedure is advisable for larger wings.

First, cut the wing planform from fairly thick plasticard. Ignore any dihedral break at this stage. If the wing is basically straight taper, cement two plasticard ribs at each end of the wing. More complex shapes will require more ribs to define

SCRATCH BUILT!

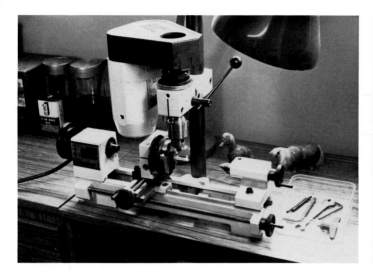

Peter Cooke's Unimat 3 can perform almost any machining operation that a modeller could require. A combination lathe, drill and mill, it is almost indispensable for generating many precision components, especially those requiring turned or radially indexed features. We offer no explanation for those critters hanging about. (Cooke photo)

The master forms for the wheels were turned using Cooke's Unimat 3. The silicon master molds and final epoxy casting of each half are also shown. (Cooke photo)

their cross section. Fill the spaces between the ribs with balsa blocks. Stick a large sheet of sandpaper to a convenient table using double-sided tape. The wing can be rubbed quite vigorously over the sandpaper in the secure knowledge that the scraping of the plasticard ribs will warn when sanding should stop, or the angle be altered. The balsa wing should now be painted with polyester resin and finely smoothed down, like the balsa fuselage.

If there is a dihedral break, the balsa should be sawn down on each side of the wing as far as the plasticard core. The dihedral angle can now be set by inserting a balsa wedge in the saw cut, and any remaining gap filled with Araldite or Plastic Padding and finally smoothed over. The ailerons and, if necessary, the wheel wells should now be cut out of the wing using a razor saw. The upper surface of the wheel wells can usually be skinned over with a single sheet, so loss of the wing shape in this area is not critical.

The wing could be skinned directly and built into the model, though a resin copy made via a plaster cast is stronger, and necessary for production runs. If the wheel well has been incorporated, a plaster cast will have to be made in two pieces. In this case, cover one side of the wing with double-sided tape and stick this down to a sheet of cling-film a couple of inches larger than the wing all around. Pull up the cling-film to form a watertight tray, as with the fuselage, and pour in the plaster mix. When set, turn the cast over and pull the cling-film into a tray in the opposite direction and pour the other half of the plaster cast. When the cling-film and wing are removed, the two halves of the plaster cast should fit sufficiently tightly when held together with rubber bands for no seepage of polyester resin to occur, as long as it is a fairly stiff mix. The mould can usually be filled via the wing root. The resulting wing casting will be too thick by about 4 thou. due to the double-sided tape. This does not matter, as it can easily be lost by final sanding.

Walls can now be built into the wheel wells using ten thou. plasticard, with the riveting detail previously impressed into them. Where the wheel well is fairly large, it may be advisable to reinforce the upper surface with thick gauge plasticard to avoid sag in the final skinning.

SKINNING: When contemplating the pattern of skinning on an aircraft you wish to model, remember that the problems you face are similar to those which the manufacturer of the full-sized aircraft had to solve. First decide which panels have double curvature, and will therefore have to be heat formed, and which can simply be curved out of flat sheet. As well as being more realistic, use of separate panels is usually the easiest way to skin a model, although several panels can sometimes be combined and formed in one piece if realistic lap-jointing is not going to be attempted. In this case the panel joins can simply be represented by scribing after forming. This is easier if done against a flexible straight edge made from a strip of plasticard stuck to the panel with double-sided tape.

I used to form panels from ten or 15 thou. plasticard. This had to be rubbed down and polished smooth to produce a realistic representation of alloy skinning, so I now use the same 10 thou. P.V.C. sheet that I use for making transparencies. This is not only smooth, but is also fairly hard and scratch-resistant. It may seem obvious to state that the final paint finish can only be as good as the surface underneath, but it is a lesson I had to learn the hard way. Paint may enhance a blemish, but will certainly never cover it up.

While vacuforming can be employed, most skinning panels can be formed by simply holding the opposite ends of the P.V.C. sheet in clamps made from double strips of balsa with a sellotape hinge. A fair degree of force often has to be used in forming, and the P.V.C. sheet will tend to pull out of the clamps unless it is attached to the inside with a strip of double-sided tape. The material, held in the clamps, should be heated in front of a fire until the sheet quivers when shaken. This is the point at which the sheet is soft enough to form well, but not so hot that it will over stretch. There is a tendency to feel that the actual forming must be accomplished quickly, before the

Chapter V: Resin Casting

sheet cools down. In practice, a slow steady pull works best. Always make at least one spare for each panel formed. Also make sure that the sheet is oversize by a reasonable amount to allow for trimming.

In the case of most wings, only the leading edge and wingtip panels will need to be formed. The rest of the wing skinning can usually be cut from a single sheet. Since adjacent panels are usually butt-jointed, the panel joins can simply be represented by scribing with a pin. After forming the wing tip in P.V.C. sheet, any navigation lamp housing can be cut away from the wing core. Unless you possess genuine vacuforming facilities, moulding inside curves like the base of a fin can be difficult without wrinkling. One method is to tape the P.V.C. sheet to one side, and then, after heating, ease it round the curve with a firmly held tissue.

Rivet Detail: The tool I use for this is a watch cog built into the handle of an old dressmaker's pattern-marking wheel. Run this alongside a plasticard straight edge using moderate pressure (a steel straight edge will chew up the brass teeth of the cog), and a neat row of "flush rivets" are impressed into the panel. On the reverse side is a row of mushroom rivets. In the case of plasticard, these tend to look more like pimples, but P.V.C. sheet will produce crisp mushrooms. Don't worry if lines of rivets are not always perfectly straight; they aren't on full-size aircraft either. When thin alloy sheet is rivetted, the material "quilts" as a result of the stress, the degree of quilting depending on the gauge of alloy. This quilting will occur on the model panel if the "rivetting" is done on a wood surface, the degree of quilting depending on the hardness of the wood.

Panel attachment to the forms: Panels can be fixed into place with double-sided tape, but I prefer to seal the edges by allowing Superglue to seep under and quickly wipe away the surplus, before there is time for the outer surface of the panel to be attacked. Panels should be trimmed and matched into place to form a close fit. This is easier to achieve with transparent skinning. When panels are lap-jointed, they are rarely thicker than two mm. and this is equivalent to thin sellotape in 1/24th scale. A thin strip of this sellotape should be applied to the underside of the appropriate panel edge to raise it slightly above the adjacent panel.

If necessary, panel joins can be finally made neater by rubbing in fine Surface Polyfilla. When set, the surplus can be wiped off with a damp tissue.

Finishing Touches: Small fittings for the fuselage can either be moulded or cast in resin using a rubber mould, as described in the next section. When making the master-patterns for items like air-scoops, radiators, or cockpit canopies, I generally define the shape with a plasticard skeleton and then fill in with Plastic Padding. After smoothing down, a resin copy can be cast. This is essential for a cockpit canopy master, as it has to be completely smooth and flawless for a really clear transparency to be moulded.

Small wing-root fairings can be made by defining the edges of the fairing with plasticard or sellotape, and then contouring in between with Fine Surface Polyfilla. This is an ideal material, as it can be quickly worked and polished to a smooth surface. Larger fairings are best made by separate panels in imitation of full-size practice.

MASTER MOULD AND RESIN CAST COMPONENT: At this point, the major "subassemblies" are represented by the master forms, covered by their rivet and panel detailed skin, with certain projecting features affixed.

This is the master form for Cooke's 1/24th scale Lancaster. The heat formed, 10 thou PVC panels, appropriately rivet embossed, have been attached to the solid polyester resin form. (Cooke photo)

Here is the epoxy cast fuselage shell. A master silicone mold had been cast from the master form, shown at left. A core piece (basically a master form, sans skin) was carefully located within the mold. The epoxy was slowly poured into the mold and . . . voila! To achieve a shell thickness of (about) 40 thou, the master form can be sized for two skins: an inner 30 thou substrate beneath the 10 thou panels. A duplicate form can then serve as the core: unskinned except for 30 thou pieces attached only at transparency locations. (Cooke photo)

SCRATCH BUILT!

DeHAVILLAND MOSQUITO Mk.IV: This closeup represents DZ353, coded GB-E of No. 105 Squadron, RAF, 1942. (Cooke photo)

DeHAVILLAND MOSQUITO B.XX: No. 139 Squadron, RAF. (Cooke photo)

Cooke adds interior detail, finish paints and installs many transparencies prior to final assembly of the major components. (Cooke photo)

Peter Cooke epoxy-cast these transparencies from a 10 thou master "skin", over which 5 thou framing was applied with superglue. The casting core was identical with the form over which the master "skin" elements were pulled. (Cooke photo)

Silicone Mould: A master mould is now prepared for each major component. Silicone rubber is used to capture the surface detail while preserving the resiliency to permit removal of the form and subsequent final part.

Clearly, forethought is required to ensure that the master form and finish casting can be extracted from the mould without damage to either – and that interior cavities can be represented. Often, the best way to achieve this, as well as to work with manageably sized "subassemblies" is to cast the fuselage (or whatever) as multiple components, which will be joined at final assembly. Consideration must be given not only to the structural logic of major component division, but also to minimizing the difficulty of later "healing" of the joint: thus, an obvious division would be at a panel line.

Almost any appropriately sized and shaped container can serve as the cavity into which the silicone is poured, so long as it can be separated after the material has cured. Since silicone rubber is expensive, one should be judicious in selecting enough wall thickness to ensure dimensional stability, but not so much as to be wasteful.

CASTING THE PRODUCTION COMPONENT: With the master form removed from the silicone mould, we are at last ready to consider subassembly production!

Interior cavities are of course created by insertion of core moulds, whose extraction from the final casting must be addressed. Various exotic approaches can be envisioned, including "lost wax" casting. But, usually the core mould (cavity) can be built up from plasticard, with inverse ("female") features added for interior detail, care being taken to assure later removal from the rigid casting. A variation here is to provide a relatively featureless cavity, whose detail is added later.

The core mould must be carefully inserted within the cavity of the silicone mould to ensure its correct relative position, even as the epoxy is being poured. This requires judicious placement of standoffs on the core, which will neither

Chapter V: Resin Casting 75

SPITFIRE L.F. IX E (TE 566): This model is finished in postwar Czechoslovakian Air Force livery. (Cooke photo)

SPITFIRE Mk.IX: This model represents an aircraft flown by Wing Commander Johnnie Johnson. (Cooke photo)

SCRATCH BUILT!

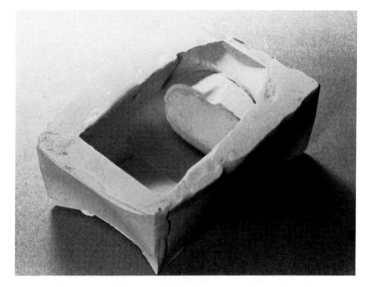

From this was cast the radiator intake housing of Cooke's Hawker Tempest. Below: Mold and epoxy cast propeller blade. (Cooke photos)

disfigure the final exterior surface nor obstruct core removal (they can be lightly attached for later breakaway from the core).

Finally, with the silicone mould back in its upright container for support, the core(s) securely in place and ample mould release applied all around, the epoxy resin is admitted, slowly and evenly so as not to produce voids (trapped air bubbles).

ENGINES, COWLINGS, AND PROPELLERS

ENGINES: At first sight some engines can appear to be of daunting complexity, but remember that most technical objects are basically collections of rectangular or cylindrical shapes. This simplicity is only obscured when the shapes are faired into each other and "bits and pieces" hung on the outside. Also, we only have to reproduce the correct appearance of the engine, and not its function. I will describe a variety of engines by way of illustration.

The Rolls Royce Griffon: As installed in my Spitfire Mk.XIV, not much of the engine is visible. The basic engine was therefore simply constructed as a rectangular plasticard box, with the visible reduction-drive casing on the front made from flat plasticard. "Boltheads" are small blobs of glue. The hidden supercharger at the back of the engine is a plasticard box surrounded by another rectangular box to represent the intercooler. All these forms were then faired in to each other with Plastic Padding. The prominent ducting to the intercooler was modelled in Plastic Padding, the initial shape being defined by plasticard "walls" which were removed once the resin had set. The cam covers were cast in Plastic Padding by filling a plasticard box. This box, of the correct trapezoidal section was held together with sellotape, and could be dismantled once the Plastic Padding had set. I decided to cast the cam covers, because they needed to be rounded considerably, and this can be done easily and smoothly with Plastic Padding.

The magentos in the front of the cylinder banks were made from plastic rodding and plasticard, faired in with Plastic Padding. The rectangular suppressor box on the engine side frame was a piece of 40 thou plasticard surmounted with a piece of 5 thou plasticard on which bolt head and rib detail had been impressed from behind with a rounded compass point. All the rest of the pipework and ignition harness was from plastic rodding and stretched sprue. Rubber pipe junctions were strips of sellotape wrapped round, with

hose clips as narrower strips of sellotape. The engine side frames were constructed in the same way as the suppressor box, but the bearing boss was turned in brass to reproduce the appearance of the original.

The Napier Sabre: This engine is basically a rectangular plasticard box, to the sides of which 4 identical cylinder heads are stuck. This cylinder head was constructed as a smaller plasticard box, open at the back, with the detail added using scraps of plasticard. "Boltheads" were small lengths of plastic rodding stuck into pre-drilled holes. The open back of the box was stuck to a plasticard base using double-sided tape. A sellotape "wall" was constructed around this base and catalyzed silicone rubber poured in. Once set, resin copies of the cylinder head could be cast. This useful casting technique can be used on any master pattern that has a flat surface somewhere, which can form the open end of a mould. The snail shell-like supercharger casing, and ducting, was constructed from heated and coiled plastic tubing, faired in with plastic padding. The characteristic long bolts holding the two halves of the supercharger casing together are lengths of plastic rodding. The air-cooled compressor on top of the engine is discs of 5 thou plasticard of alternate diameter from a leather punch stuck together with a tiny dab of Araldite. Apart from turning with a lathe this is often the only way to produce true scale finning for such things as air-cooled cylinders, even in 1/24th scale.

The Mareng type fuel tanks were cast in plastic padding using a plasticard mould in which the "fabric covered rivets" had been impressed with a rounded compass point. The edges of the tanks were rounded off afterwards.

Chapter V: Resin Casting

The Bristol Centaurus: The 18 cylinder radial for the Sea Fury is shown in the accompanying figure. Even on close examination of the actual engine, nothing beyond the cylinder heads is visible in this compact installation. The cylinder heads were thus mounted on a blank drum. This can be fabricated out of plasticard with the join out-of-sight underneath, or turned on a lathe from perspex. In the latter case the cylinder head positions can be marked or drilled using a dividing-head. In the former case they are best marked out on the plasticard before bending up. The mounting drum is invisible if painted gloss black, rather than matt black.

This engine is a classic case for casting components in resin. The apparently complex engine is assembled from multiple castings of just four components: cylinder head, air guiding shroud, left and right handed exhaust pipes. The ignition harness is fine plastic rodding.

A large engine-cowl like the one on the Sea Fury cannot be adequately represented in 1/24th scale by conventional heat forming, as the rounded lip and internal shape can be seen. This was therefore turned on the lathe as a hollow ring with a solid back wall, and a resin copy made using a silicone rubber mould. The Q3-3481 rubber is sufficiently flexible to pull free from the internal undercut. The solid back wall formed the open end of the mould.

COWLINGS: I will describe construction of a Tempest cowl in detail since, although fairly complex, it illustrates principles of wide application.

The shape is too complex to be heat formed, and so must be cast. A master-pattern is constructed in which the shape is defined by a plasticard skeleton. Any apertures in the finished casting are simply represented as slight recesses at this stage. It is rare for shapes like this to be completely defined by drawings. In this case each plasticard frame was taken directly from one of the sections shown in the excellent plans by Arthur Bentley (M.A.P. Plan Pack 2943). The gaps in the skeleton were then filled with plastic padding and roughly contoured with sellotape. After the sellotape had been removed, the shell was sanded until the plasticard frame lines were just visible everywhere.

This is a good method for producing cockpit canopy masters, but the plasticard frame lines cannot be completely obliterated so a copy will have to be cast in resin for final polishing.

The air-intake duct gets wider as it goes in, and had to be produced as a separate master-pattern. This is partly because it is a shape easier to produce as a "male" pattern, and partly because the stiff rubber I was then using for moulding would not have pulled through the front opening. The rubber copy of this "male" duct then had to be produced in a two-stage process, casting silicone rubber in a silicone rubber mould with Vaseline as a release agent. The two rubber moulds were pinned together as shown in the accompanying figure.

Since I needed a thin shell as on the full-size aircraft, the casting was produced as a "slush moulding," i.e., thickened polyester resin was swilled into all the recesses of the mould until it had started to "gel." When the resin was hard the moulds were unpinned and then removed separately from the casting. The radiator itself was constructed from plasticard, the "radiator matrix" being cross-hatched with a scribing pin. The truncated cones of the inner duct were moulded in the same way as spinners, and cut to length after moulding.

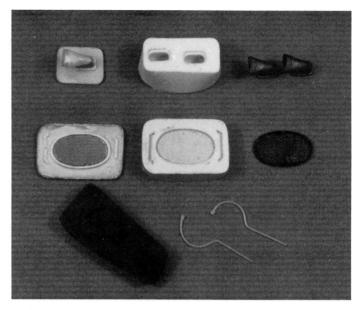

Cooke tidbits. (Cooke photo)

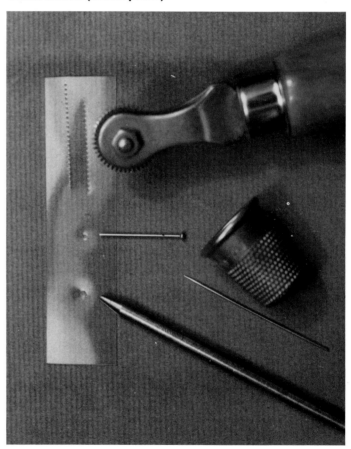

The holder mounted watch cog is of course used for producing rivet lines in clear 10 thou PVC.

EXHAUST PIPES: Multiple exhaust pipes can always be reproduced in resin with their mounting faces as the open end of a mould. The silicone rubber is sufficiently flexible to allow withdrawal of even fishtail types. These tiny moulds must be filled slowly to prevent air bubbles getting trapped. I suspect

SCRATCH BUILT!

This composition of Hawker Sea Fury components speaks eloquently of Peter Cooke's resin cast modelling achievement. (Cooke photo)

Further visual testimony to the effectiveness of the Cooke approach. Note that the resin cast landing gear legs include an embedded steel reinforcing element. (Cooke photo)

that these pipes were fabricated as cylinders and then flattened at each end at right-angles. This is certainly what I did, using a thin plastic drinking straw and automatically got the right shape. The mounting end of the pipe was built up in plasticard and faired in with plastic padding. These plastic straws, or biro refills, can be stretched by heating to reduce their diameter if necessary.

PROPELLER BLADES: It is a mistaken notion that modelling should be done with "authentic" materials, since these rarely look correct when scaled down. One exception occurs with the earlier wooden type of propeller blades. There is no better method than to cut these blades from a glued lamination of marquetry veneers (thinned down in the smaller scales). After shaping, and finally sanding with thousand grade paper, these look incomparable if wax polished, or varnished. The veneers should be chosen to approximate in colour to the wood of the prop being modelled.

One problem with shaping propeller blades is that draw-

Chapter V: Resin Casting 79

SPITFIRE Mk.IXC: This model represents MA454, UM-V of No. 152 Squadron, Italy, 1943. (Cooke photo)

SPITFIRE Mk.IX: This is a model of Stephen Gray's ML417 in its true wartime guise, with No. 443 Squadron, RCAF, in autumn 1944. Model was built for the well known aviation painter Robert Taylor. (As opposed to our movie star – J.S.A.) (Cooke photo)

ings rarely give enough information. These can be drawn from carefully taken photographs. 35 m.m. slides can be projected to the required scale and traced. It is not always easy to get the camera at right-angles to the leading face of a blade when the aircraft is sitting on its tailwheel, which is why I carry a small step ladder in the car on such trips, although this does cause a few raised eyebrows occasionally. These drawings, supplemented by a few key dimensions, will enable the developed, i.e., "untwisted" shape of the blade to be drawn. This is then cut from fairly thick plasticard with a gap to allow for the centre spigot. This spigot is turned in epoxy resin or brass using a lathe, or an electric drill, and includes the cross-sectional shape of the blade, as well as the mounting boss. This not only helps the final shaping of the blade but will ensure that the blade is not bent during assembly. The plasticard blank is now twisted to shape, checking against the front and side views and then Superglued to the spigot. Plastic padding completes the contouring of the blade. Even in 1/24th scale I find it is not possible to produce a true scale thickness towards the tip of the blade without the plasticard becoming very brittle, so no attempt is made to thin the plasticard at the tip, or along the edges at this stage. Instead, a rubber mould is made by mounting the blade upright on a small base using double-sided tape, and then building a tall wall around it with cardboard or plasticard. The rubber must be poured in slowly to prevent air bubbles getting trapped. An epoxy resin copy of the blade is then cast, squeezing the mould to expel air bubbles. This is then honed to a really fine profile, and a fresh rubber mould produced, from which as many resin copies can be cast as is required.

The prop blades can be mounted by glueing the moulded spinner, complete with any rivetting detail, back onto its master-pattern and then drilling this to receive the mounting bosses of the blades. If you do not want to lose your master-pattern, make a resin copy, complete with skinning and rivet detail, prior to drilling the blade apertures. Epoxy resin drills and machines extremely well.

PAINTING

Since my technique for painting resin cast models is very different from that usually employed for plastic models, I will explain it here. My motivation for developing this approach was to replicate as well as possible the actual silky sheen appearance of wartime camouflage on metal skinned surfaces, whilst preserving the intricate rivet/panel line detail achievable with PVC skin/resin cast construction.

PAINT CONSTITUENTS: By a long trial and error process, I arrived at the following rather bizarre formula:

- Mix matte HUMBROL from the tinlet with an equal quantity of enamel thinner;
- Add about the same quantity of polyurethane varnish (or perhaps slightly less);
- Add somewhat more nitrocellulose thinner than for either of the other two constituents. I now mix until just a trace of color remains on the side of the container when touched by the mixing stick.

Polyurethane provides the correct sheen, tightens as it dries and imparts a hard surface. It can later be rubbed down if necessary, with worn 1200 grade wet or dry. The nitrocellulose thinner ensures smooth flow through the airbrush and makes the paint dry more quickly. It seems to prevent coagulation of the mixed paint.

APPLICATION: Along with the aforementioned formula, the second key to achieving the desired result is a very thin air brush application of paint – only enough to impart opacity. Otherwise, the fine rivet and panel line indentations would become flooded.

But, no finish is better than the surface upon which it is laid. I have found that clear P.V.C. sheet possesses an ideal surface for subsequent moulding and casting. All that is required is a brisk rub down with a clean cloth. Of course, the parent sheet must be carefully examined for surface flaws before use. (Note: This is in stark contrast to the usual necessity for extensive sanding of the vacuformed polystyrene sheet, in order to achieve the desired crispness of shape, as described in Chapter IV – Alcorn).

THE FINAL SURFACE: After two days, the polyurethane varnish will be dry, but not fully hard. At this point, black water color can be rubbed into the surface. When most of it has been wiped off, an oily "used" appearance is achieved. The black will also remain in panel lines and around rivet heads, highlighting those features.

I do not like oversprayed varnish, since it always seems to give the paint an unrealistic depth, although I can see its advantage for hiding decal film. In any case, I airbrush all of my insignia/markings directly onto the surface, using masks cut from low tack frisket film.

CHAPTER VI: DETAILS

INTRODUCTION

At this point in our "how to" narrative, we are beginning to merge into the realm of the kit modeller, at least those of the "superdetail" and conversionist (revisionist running dog) persuasions. By Chapter VIII: PAINTING, you won't be able to tell us apart, although we plan to stage a comeback in Chapter IX when we will cover silk screen techniques for producing "scratch-built" multicolor decals.

Close-up of George Lee's Keystone.

SCRATCH BUILT!

Here is a sample sheet of the pen and ink (Rapidograph on Bristol Board) patterns which I prepared (to 4x scale) for my 1/24th scale DeHavilland DH9A. In addition to this, I prepared another 4x sheet, a 6x sheet and a 2x sheet, each original being about 10" x 14". The originals were sent to Fotocut for photoreduction and etching. Various brass thicknesses were required, from 2 to 5 mils. Sorry about the inadvertant "Smiley Face." (Alcorn photo)

Lest we be maligned as a cliquish bunch of modelling snobs, let us assure you, dear reader, that we in no way look down our noses upon our kit bashing brethren. We've never been to a regional or national IPMS event yet where we didn't stand in awe before some exquisitely rendered kit model, often the handiwork of some guy half our age, or less. Often it is artistry of finish, whether it be simulated natural metal, weathering effect or complex coloring and markings executed to perfection. Perhaps it is a *tour de force* of interior detail, or a superb conversion. Or maybe it is just a generally outstanding job on some aircraft which we particularly admire.

Since "details" means everything other than the basic structure for all types of aircraft in a 50-year time span, we're implying a vast subject, to which we couldn't possibly do justice in a single tome such as this. Besides, no three modellers could possibly know all the tricks. So, we'll defer to the many fine magazine articles and other modelling books for countless esoteric techniques, such as making spoked wheels and photo-etched gun barrels. Instead, we'll concentrate upon those methods with which we are familiar, or refer to others.

Even by its "legal" (i.e., IPMS) definition, "scratch-built" doesn't imply that every component be "cut from whole cloth." To be sure, the basic structure must be, but from there on you must let your conscience, and expediency, be your guide. After all, it would be a bit silly to construct Pratt and Whitney R-1860 engines for your Keystone B-4A when you can modify them from a Hasagawa Boeing F4B-4's R-1340.

COCKPIT/VISIBLE COMPARTMENTS

We're now firmly within the province of the superdetailer, who lavishes most of his overachieving psychosis into interior detail, some of which is too small to be appreciated with the naked eye – and some of which can't be seen at all once the model is complete! – but he knows it's there and that's what matters, right? Weird.

Chapter VI: Details

But, cockpits are important, and satisfying to see. Notice how much of a contest judges' time is spent squinting intently into the bowels of the subject under scrutiny, whose intimate secrets are revealed by the merciless beam of his penlight – as its creator stands by terrified both by what he'll find and by the possibility of the light becoming a missile, plowing its way through delicate incrustations of minute levers, handles, and wires.

INTERNAL STRUCTURE: We've already touched upon visible interior structure which is made integral with the fuselage shells –longerons, stringers, braces, keel beams, etc. Often, however, additional structure must be added which is more or less independent of, or superimposed upon the integral shell elements. An example is the tubular structure visible within the cockpit of a Hawker Hurricane. The side trusses can be made from brass tube, soldered together and installed before joining of the the shell halves. The cross braces are then added using epoxy, applied sparingly with a pointed stick. The exact length of each cross brace can only be approximated, except by trial and error.

INSTRUMENT PANELS: Waldron Products markets a line of printed sheets for U.S., British, German and Japanese WWII vintage instrument dials, to 1/48th, 1/32nd, and 1/24th scale. Prior to the advent of these fine products, I (Lee) achieved similar results by obtaining scale photoreductions of actual (U.S.) dial faces. Positives on glossy paper were obtained, which could be individually cut out and glued to the backside of appropriate holes in the dash to be simulated.

A typical WWII vintage panel would be made as follows: Very carefully lay out the dial center locations upon a polystyrene sheet blank, of say 30 mil. This is an Optivisor job, even for the sharpest of young eyes. Using a bulletin board push pin, prick each location. With a vernier compass, lightly scribe each dial hole circumference. Then, starting with very small drills (say #80) mounted in a pin vise, gradually increase the holes until the final sizes are approached. From then on, making position corrections as necessary, open the holes to full size using a rat-tailed Swiss file.

Usually, circular bezels are required for each, or most dials. These tiny rings can be cut from 10 to 20 mil sheet by successive, progressively deeper scribing with a vernier compass, deftly spun with the fingers. As can well be imagined, this is a very tedious process, demanding great patience and accuracy. Once made, these tiny rings are carefully MEK'd in place and cleaned up with No. 600 sandpaper.

The panel is air-brush painted and the dial facsimiles glued from the backside using Elmer's (it dries clear, doesn't effect the plastic and allows for trial and error dial positioning – as viewed from the panel front). Typically, numerous other features must later be added to the panel, including levers, knobs and handles.

PILOT'S SEATS: In his book *Scale Model Aircraft*, Harry Woodman briefly illustrates and discusses the basic types of seats most commonly found in aircraft of the 1915-1945 era, though he demurred at explaining construction of a 1/48th scale "wicker lawn chair." While many seats can be built up from sheet and scrap stock, those metal examples with flared edges and parachute wells are often best vacuformed over carved forms.

INTERIOR DETAILS: Due to the enormous variety of smaller objects found within cockpits and other visible crew areas, it is impossible to provide a comprehensive item-by-item discussion of their representation within models. These include seat belts and cushions, throttle and mixture controls, separate instruments, rudder pedals, control column, trim tab controls, bombsights, cameras, gun rings and mounts, ammo boxes, control cable runs, first aid and survival kits, fire extinguishers, radios, radar consoles, reflector gunsights, chart tables, lights, oxygen bottles, canopy hatches, electrical panels and relief tubes. One need only to contemplate the interior of a B-17G to experience either the rush of challenge or the crush of despair.

The extent, authenticity, and neatness of cockpit detailing is limited only by the modeller's skill, patience and imagination, as well as by the richness of his "bits box." Prototype elements for smaller parts can come from such diverse sources as plastic aircraft kits, ship and train model accessories, electronics stores and the wife's sewing box. Among the most useful raw materials for interior detail are the following:

- POLYSTYRENE STRIP: Evergreen Scale models of Kirkland, WA markets a line of crisp rectangular polystyrene strips in 13" long packets, which is extremely useful for interior structure and detail. For those of us who have been cutting our own from sheet stock for many years, this product is a godsend. Slaters of England markets a similar line, aimed specifically at the steam and whistle crowd.
- FINE WIRE: Copper, aluminum and steel in the 0.005-0.025 inch diameter range. Such wire is available from hardware and electronic/electrical stores.
- THIN POLYSTYRENE SHEET (5 mil): This is available in packets by Evergreen and others from the more complete hobby shops.
- THIN ALUMINUM SHEET (5 mil): The best source for this material is cast-off photo plate sheet.
- BRASS SHIM STOCK: K & S Engineering Co. of Chicago packages this indispensible material in small sheets as thin as 1 mil.
- POLYSTYRENE SPRUE: This we, of course, make ourselves by the traditional heat and pull technique.
- PLASTIC ROD: This is available in assorted small diameters from better hobby shops. For some reason, it is usually brownish in color.
- PLASTIC STRUCTURAL SHAPES: Plastruct markets a line of scale "I" beams, "H" beams, "T"s, channels and "angle irons" for architectural modelling, which can be useful.
- TELESCOPING TUBING: This invaluable material, supplied by K & S, is available in brass and aluminum, from 1/16" to 1/2" diameter from most any reputable hobby store.
- FINE BRASS TUBING: Picking up where K & S left off, Hobby Hanger (P.O. Box 472, New Carlisle, IN 46552) markets an amazing line of 6" long brass "Micro Tube" in 57 sizes, ranging from 8 to 63 thou outside diameter. It's one of those products we don't know how we ever lived without before.
- FINE STAINLESS STEEL TUBING: Accurate Detailing Inc. of Aurora, CO, markets a line of medium hardness stainless steel "Detail Tube"; in packets containing three each 2" lengths; in eighteen sizes, varying from 0.006" O.D. x 0.002" I.D. to 0.050" O.D. x 0.033" I.D.

SCRATCH BUILT!

VICKERS C.O.W. – FIGHTER: Trust the Brits to build this contraption, and George Lee to immortalize it in plastic! The layout of the Vickers 161 was driven by the need for unrestricted forward operation of its Coventry Ordnance Works 37 mm rapid firing, manually operated cannon. Its most bizarre feature is the fuselage extension beyond the pusher mounted Bristol Jupiter engine. Doubtless only its creators could explain why they didn't simply rely upon the twin booms for tail support. Needless to say, the type was not adopted for service. (George Lee photo)

WESTLAND PTERODACTYL Mk.V: Brilliant concept, or misbegotten sport? This sesquiplane flying wing was meant to feature an electrically operated turret just behind the pilot. While capable of reasonable performance for 1932, the Rolls-Royce Goshawk powered Pterodactyl was not the way to fly, at least for interceptors. (George Lee photo)

Arlo Schroeder constructed this 1/16th scale Avenger for the NASM. It represents a machine of VT-10, with which he served two tours as a turret gunner; aboard the Enterprise (CV-6) and later the Intrepid (CV-11). (Photos courtesy Arlo Schroeder)

• **EPOXY:** Usually of the two component "5 minute" variety, it is indispensable for producing tiny lever knobs and assorted "lumps" as well as for certain joining jobs.

Photoetching: In recent years, the "state-of-the-art" of interior (and exterior) detail has been considerably enhanced by the advent of photoetched parts – "fallout" technology from the electronics industry. Incidentally, Harry Woodman avers that in fact he introduced photoetching to the aircraft modelling community in general. Details of his technique were published in the November 1972 Scale Models. In 1975, he produced designs for fretted gun jackets, spoked wheels, basket seats, propeller bosses, bomb carriers, sights, bezels, etc., which were featured in the September 1976 Scale Models, and marketed in 1/48th scale by Fotocut, then of Canada. Many exquisite photo etched parts are now available from several specialty manufacturer's. Cockpit interior sets are the usual fare for aircraft: many exotic components for ships, cars and

Chapter VI: Details

These panels were made by Peter Cooke, using the technique described herein. From top left: Lancaster bomb aimer's; Spitfire Mk.IX; Sea Fury; Lancaster engines and pilots panel; Mosquito Mk.IV panel, plus compass. (Cooke photo)

Right: More Cooke resin cast components, these are for a Spitfire Mk.IX. (Cooke photo)

trains should not be overlooked. Representative suppliers include Tom's Modelworks of Cupertino, CA, and Model Technologies of Chino, CA, as well as Fotocut.

Scratch builders can prepare their own black and white "artwork" (to 4x, or greater scale); providing 1:1 scale photo negative transparencies to job shops specializing in photo etching for the electronics industry. (see photo page, 82)

GUNS

As regards machine guns, you're fortunate if you can locate a useable example from a model kit, or from vehicle/figure accessories. However, even the best of them usually require upgrading to scratch-built standard.

The most challenging part of a machine gun is the fretted jacket which surrounds the barrel of many vintage types. The WWI Spandau, Parabellum, Schwarzlose and Maxim examples were modified from water-cooled infantry weapons, the jacket being fretted with elongate slots for air cooling. While flat photoetched brass stock is available, we have been unable to satisfactorily "roll our own." So, despite the pain, we've resorted to marking, drilling and filing out frets in a piece of telescoping tubing – usually thinned down by reaming the insides with pin vise-held drills. The most crucial part is

86   SCRATCH BUILT!

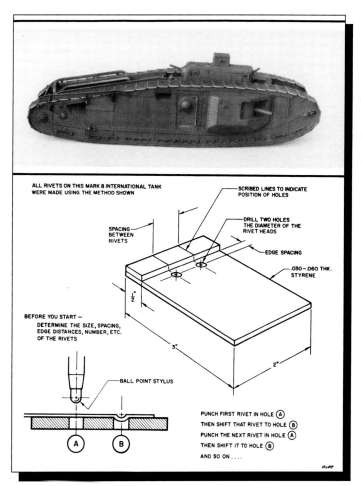

Making large rivet heads. (George Lee illustrations.)

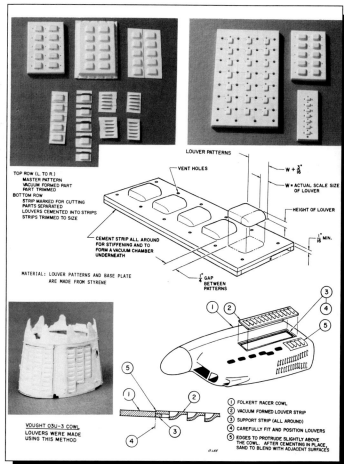

Above: Making louvres. Below: Corrugated panels.

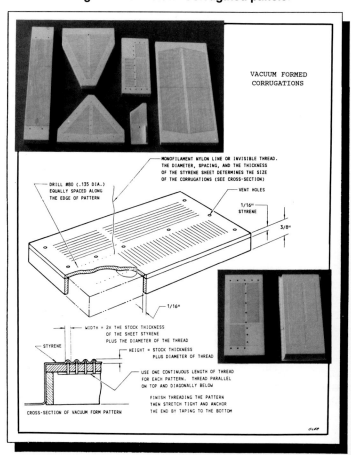

accurately laying out the hole center locations and "center punching" them with a bulletin board push pin. Opening the holes with successively larger drills is boring (pun intended). Frets can be produced, with difficulty, by elongating holes with a round Swiss file. Incidentally, in some cases, it is necessary to thin the aluminum tube from the outside, using wrapped sandpaper. The smaller the barrel housing diameter, the worse it gets: WWII 0.30 and 0.50 caliber barrels to 1/48th scale being the ultimate.

By comparison, the rest of the gun: body, surface features, hand grip, trigger, ammo feed and mounting is a "piece of cake." Whether installed within turrets, or manually operated in open mounts, it is usually best to allow the guns to articulate. This permits them to later be positioned to best suit the most realistic "look" of the position.

Another severe challenge for the dedicated "gunsmith" is to produce crisp, neat, even, scale belted ammunition. Again, photoetched parts offer a solution.

LANDING GEAR

This is a major topic of itself, given the enormous variety of such components within the scope of this tome. Generally speaking, earlier fixed landing gear is easier to model than the intricate, articulated, oleo strutted mechanisms of retractable gear.

Chapter VI: Details

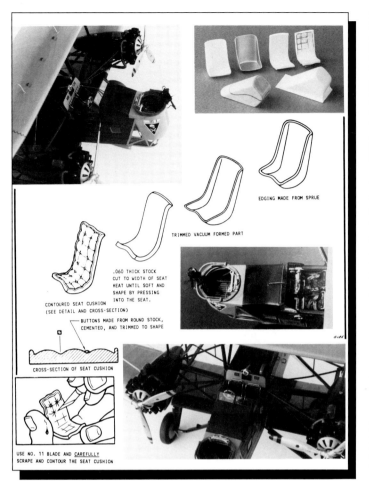

Making seat cushions. (George Lee illustrations.)

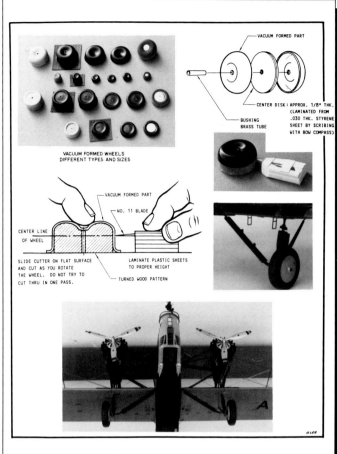

Making wheels.

RETRACTABLE LANDING GEAR: For these assemblies, the more complex portion is usually located within the wheel well, wherein reside anchor points, substructures, hydraulic cylinders, springs, latches, clevises, swivels and myriad other modelling horrors. To appreciate the enormity of the task, reflect for a moment upon the mechanism of a Grumman F4F Wildcat. There are even those of a micromechanical bent who would consider producing retraction mechanisms from metal which actually work. To us though, this smacks of sadomasochistic phobia, since it violates the concept of "static scale."

Even for static building though, metal is usually best for the primary support elements, soldered together as appropriate. Telescoping brass tubing is of course ideally suited for oleo strut simulation, collars being fashioned from thinned short lengths – "superglued" in place. Heresy though it may be, for some gear to aircraft structure attachments it may be prudent to discreetly anchor the sturdy leg into a rigid hard point, rather than through flimsy scale fittings.

Soldering, brazing and welding of metal are usually considered specialist skills, known only to professional technicians, or modernistic sculptors. In fact however, soldering, at least for our limited purposes, is really quite basic as regards skills and equipment and *de rigour* for strong landing gear, or other metallic space frames. For equipment, all you really need is a small pencil or gun-type soldering iron, plus solder wire (50/50 or 60/40 lead/tin), paste flux, a small flux brush and some rudimentary ad hoc fixtures for holding parts. The canons of soldering are as follows:

- heat the parts to be joined, which in turn will melt the solder wire upon contact;
- contour the mating surfaces for the best possible fit;
- thoroughly clean the mating areas, which are then covered with flux to prevent oxidation during heating;
- use a minimum of solder.

As with most acquired skills, your best bet is to learn the ropes from an expert – it won't take long. For "book larnin'", we refer you to *Fine Scale Modeler*: December 1989 "The Fine Art of Soldering"; and December 1987 "Basic Soldering Techniques."

Five mil aluminum or 5-10 mil brass sheet stock is often efficacious for producing small brackets, levers, etc., associated with retracting gear. For really small, complex features, such as the scissors link of oleo legs, 5 or 10 mil styrene sheet usually yields the best facsimile.

Messier undercarts are messier (who could resist it). Certainly the "fabricated" structure of many WWII RAF bomb-

88   **SCRATCH BUILT!**

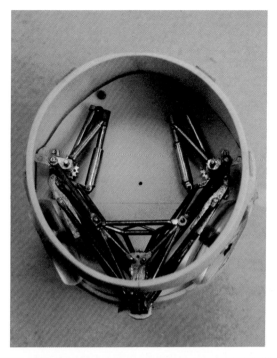
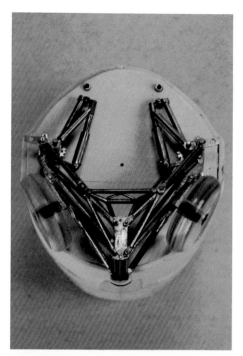

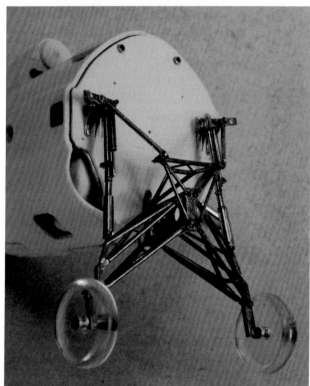

Retractable landing gear mechanism for Grumman F4 F-4.

Prophetic prose: Long after having penned the words in the accompanying text (Retractable Landing Gear, first paragraph) George Lee showed me these photos of Matt Matsushita's 1/72nd scale Wildcat! Matt provides the following detail: "Including the styrene lined brass sheet bulkhead, 110 pieces make up the assembly (which is not complete as of this writing). Most of the parts are made from brass rod and tube. Piano wire and stainless steel tubing are used where strength (main support), appearance (section of oleo strut) and size (shock cylinder) dictate. There are 16 (2 sets of 8 each) parallel hinge lines which allow the landing gears to extend and retract. The task of duplicating parts – right and left sides are interchangeable – was possible through the use of a fixture which served first as a drill jig and then as a soldering jig. The springs in the shocks are stainless steel wire (0.004"Ø) which was wound around the shank mof a 0.225"Ø drill."

Chapter VI: Details

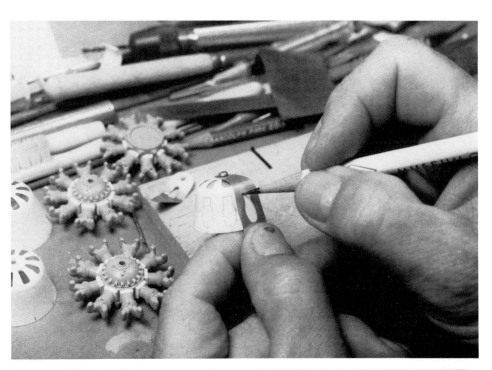

BOEING 80A-1 ENGINES: Here we see Bob Rice (well, a few fingers thereof) laying out the louvered openings of the reduction gear covers for the Pratt and Whitney Hornets of his 80A-1. He is using an aluminum template laid over the wooden vacuform element. The Hornets began life as Williams Brothers Wasps. (Rice photo)

SEATS AND FLOOR OF BOEING 80A-1 CABIN: A testament to Bob's attention to detail – even though they can barely be seen in the completed model – are those comfy-looking seats, whose legs will be inserted into the floor panel holes. (Rice photo)

ers are worthy of Goethels or Eiffel. At least they're unlikely to spontaneously collapse on the exhibition table.

WHEELS: Tom Lehrer had a song for it: "Plagiarize!" Before squandering time and energy upon scratch-built wheels, be sure that none are available from some kit – especially if a complex hub or tread pattern is required.

For thin, smooth vintage tires, rubber "O" rings are often just the thing. However, smooth balloon tires can be vacuformed fairly easily over carved hardwood "donuts" – split, of course, at their center plane. Contoured hubs can be vacuformed separately. While a central, full diameter disc could be built into the wheel, it may be better to leave it "squishy" for the flattened foot print effect.

Another technique for making wheels is to laminate solid ones from polystyrene sheet. To either side of a central disc (or discs), rings are glued in place of successively smaller O.D. and larger I.D. For my DH9A, I cut these discs and rings from 20 mil sheet by spinning a compass equipped with a sharp-pointed needle. The process was tedious and boring, resulting in very sore fingers, but the final product was fully effective, once the lamina had been carefully scraped and sanded to a circular tire cross-section. Although I haven't tried

90 SCRATCH BUILT!

BASS WOOD MASTER PATTERN - NOTE THE LAYOUT LINES
FOR THE TRANSPARENT SECTIONS

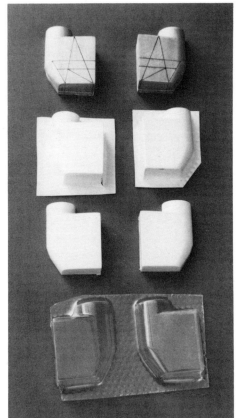

THREE DIFFERENT PARTS WERE VACUUMED FORMED
FROM THE SAME MASTER PATTERN

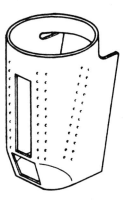

.010 THICK STYRENE

RIVETS WERE EMBOSSED
ON THE INSIDE SURFACE

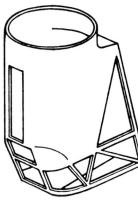

.030 THK. STYRENE

MAIN SUPPORT STRUCTURE FOR THE MODEL
WITH CUTOUTS FOR THE TRANSPARENT SECTIONS

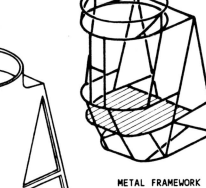

METAL FRAMEWORK
MADE FROM SPRUE, RODS,
& STYRENE STRIPS

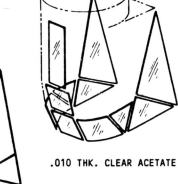

.010 THK. CLEAR ACETATE

SECTIONS TRIMMED TO FIT
THE FRAMEWORK

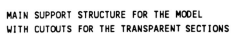

VACUUM FORMED KEYSTONE BOMBER NOSE SECTION

G LEE

(George Lee illustration.)

Chapter VI: Details

Curiously, the first Il'Ya Muromets G-3, as represented by Bob's model, used two different types of engines: the inboard pair were 220 hp Renault V-12s, while the outboards were Russian built R-BVZ-6 inline sixes developing 150 hp each. The motive was simply engine availability. The master crankcases and cylinder blocks for each engine were carved from plexiglass: the duplicate crankcases and other large components were epoxy resin cast in RTV molds. Most smaller bits were made from wire, tube, plastic rod and styrene sheet. The photoetched rocker arms were attached to the pushrods. Bob spent around 700 hours on this model, one third of it scratchbuilding the engines. I, Alcorn, am amazed that he could have built this awesome beast in such a short time. He must have been on a Rice (heh, heh) diet, sez George. (Rice photo)

Bob Rice built up this Il'Ya nose component from 10 thou clear sheet, sandwiched between photoetched brass frames. First he drew up the frame templates at 3 x scale; and test fitted paper copies over a shaped form until correct fit among the several elements was obtained. He then sent the corrected artwork, reduced to 1 x scale, to Fotocut. He assembled the nose section over the form: inner frame, glazing and outer frame – the double-curved lower corner pieces were vacuformed. Epoxy was used throughout for bonding, and subsequent attachment to the fuselage. (Rice photo)

it, circumferential tread grooves could be produced by laminating with alternately greater and lesser O.D. rings of the appropriate thickness.

ENGINES AND PROPELLERS

ENGINES: This topic was well covered by Peter Cooke in Chapter V. While it's true that resin casting is the only logical approach for producing multiple, identical components such as cylinders, many engines have been scratch built by serial replication of what could have been the master form. Again, don't overlook use of engines or components from commercial kits. In this case some rework is usually necessary, either simply for improvement (including "superdetailling") or for modification.

PROPELLERS: We include this subheading mainly for completeness since, again, resin casting and kit pilfering are primary choices. We (Lee and Alcorn) generally carve ours from basswood: if only two or three blades are required, this is not too illogical – frankly, we enjoy it.

Laminated wood propellers of the pioneer/WWI/20s eras are best produced by just that method. Select wood veneer of the appropriate thickness and color, and glue the well clamped stack with Elmers. The Axial propeller for my 1/32nd scale Rumpler CIV was produced in this manner, using basswood and walnut strips.

CHAPTER VII: ASSEMBLY

NON-RIGGED AIRCRAFT

While preparation of a jigging fixture for assembly of rigged, multiwing aircraft is essential, it is usually appropriate for unrigged monoplanes as well. The crucial assembly step is to achieve exact positioning of wings to the fuselage: dihedral, incidence, front and plan view perpendicularity. This was discussed in Chapter IV, from the standpoint of providing proper hardpoints for the wings, as a function of basic attachment method.

DIHEDRAL: How many times have you seen an otherwise nicely turned out kit model ruined by lack of adequate dihedral? The basic problem can usually be traced to kit design: a wimpy attachment point, or a tongue-in-groove (or whatever) attachment method which does not assure proper dihedral in a "fool-proof" manner. The modeller assumes that, if glued firmly together, wing to fuselage fit will be correct – that is, he has implicit faith in the kits' manufacturer. In most cases, it is necessary to carefully check dihedral before gluing: it is usually necessary to make some minor corrections to the as-injected mating surfaces. Further, on many kit models, especially larger ones, it is also appropriate to stiffen the joint with additional internal braces. Finally, all too many kit models suffer from wing droop, so that even if the root joint is correct, tip-to-tip dihedral is inadequate. This can only be remedied by adding a stiff spar within the wing. But "Hold it," you cavil, "This book is about scratch-building, not kit bashing!" Just trying to emphasize the point, man, that a rigid "fail-safe" wing attachment scheme is essential.

INCIDENCE: While rarely a problem with kit assembly, assurance of proper wing incidence – angle of attack – is a crucial aspect of scratch-built assembly. Since incidence is

Chapter VII: Assembly

BOEING 80A-1: This 1/32nd scale gem by Bob Rice represents one of twelve aircraft which served Boeing Air Transport exclusively, on a single route, San Francisco to Chicago, between 1929 and 1934. Among the 80A-1's claims to fame is the first employment of airline stewardesses, all of whom were registered nurses. In additional to their normal duties of attending to passenger comfort, nourishment, safety, and egos, the stewardesses were frequently called upon to assist with baggage handling, cabin cleanup and even refueling the aircraft. But, their lives were glamorous, exciting, and often led to marriage. As stated in his excellent September 1987 Scale Modeler article, replicating this handsome aircraft was a major challenge, due to the combination of fabric plus corrugated metal exterior, cabin/cockpit detail, fully exposed radial engines, biplane configuration, triple tail, and complex paint scheme/markings. This model enjoyed the honor of winning First in the large scratch-built aircraft category, and being voted Most Popular at the 1986 IPMS Nationals in Sacramento. It is now on permanent display at the Museum of Flight in Seattle. (Rice photo)

---

relatively difficult to produce accurately with assembly jigging, it is usually best to incorporate it into a fool-proof design of the fuselage hard point. This was discussed in Chapter IV.

PLAN VIEW PERPENDICULARITY: This is usually best achieved with a proper jigging fixture, although sometimes it is sufficient to simply measure (equalize) the diagonal distance from fuselage tail post to wing tip.

THE ASSEMBLY FIXTURE: The typical fixture consists of a flat, rigid baseboard – larger than the model – upon which are mounted various cradle, butt, resting and index jigs as appropriate to fully constrain the assembled components. These raised features are usually best made from basswood stock, reinforced with doublers and gussets to the baseboard for dimensional stability.

The fuselage is typically held in its precise fore and aft, upright position by vertical post pairs, front and rear, to which are often fitted top and bottom half-section templates, in order to prevent transverse rotation – and to hold it level in the longitudinal direction.

In a typical fixture, basswood blocks are placed at each wing root location, to assure even and precise height: they may also have a lower airfoil shape to provide added assurance of correct incidence. Blocks are set near each wing tip to set dihedral. Wing to fuselage perpendicularity is provided by reinforced vertical posts which bear against the leading edge, preferably near the tip to minimize angular error. The horizontal tail-plane can be jigged in a similar manner. Assuming that the wings are precisely aligned, the stabilizer can later be checked by sighting it to the wing, looking from the front – if one side is low despite your best jigging efforts, it will "optically" touch the wing first on its low side. Sometimes it is advisable to elevate the main structure jigging points above the baseboard sufficiently to later be able to jig the landing gear.

Once the elements to be joined are set in place, they must be firmly held down against each jig locating surface. The usual method is with stout rubber bands, pulled down over the component and anchored over nails, hammered into the baseboard at an angle to prevent the band from popping off. Often, smaller rubber bands are anchored to the jig structure using lighter nails.

It is worth noting that the jigging fixture may also be suitable for later transport of the finished model – within a suitable container. For this purpose, the landing gear should hang free – it should not be restrained in any way during shipment. I (Alcorn) learned this the hard way from commercial transport of my models from California to Virginia in 1988.

JOINTS, FILLETS AND SURFACE DEFECTS: Once the main components have been assembled, considerable work is required to fill the resulting joints, produce fillets as appropriate and to prepare the surface for paint finishing.

Seams: For seam and joint filling, we're partial to a thick paste of powdered polystyrene and MEK. After all, when dry, this is identical to the parent material. A serious pitfall to any seam/joint filler, except epoxy, is its tendency to long term shrinkage, as the trapped volatiles gradually diffuse out through the hardened surface. The best remedy for this is threefold: by minimizing the seam in the first place, by careful fit and/or insertion of plastic sheet shim stock; by application of highly viscous polystyrene/MEK paste in thin layers; by exercise of considerable patience in allowing the seam/joint to thoroughly dry before final sanding and painting: weeks is the appropriate time scale for larger jobs.

Epoxy "cures" by molecular crosslinking rather than drying by loss of volatiles – therefore, its shrinkage is essentially zero. It also adheres well to a clean, dry surface. But, unlike MEK'd plastic, it doesn't fuse with the parent material, so is more prone to eventual cracking at the epoxy – plastic interface.

Fillets: Relatively large fuselage to wing fillets, such as found on the Spitfire, Douglas DC-3 and SBD, are best vacuformed integral with the fuselage – that is, carved into the wood form. In such cases, the fillet can later be trimmed and featheredged to glue down against the wing skin at assembly. Alternatively, both wing root and fuselage can be fitted with "ribs", which mate along the fillet to wing intersection – the wing skin inboard of this region having been cut away. (Provision must have been made in the fuselage centersection to receive the wing spar(s).)

Modest wing root filleting can be produced after fuselage/wing assembly by a combination of "custom fitted" scrap styrene for the trailing edge "web"; thin strips laminated over the upper (and/or lower) joint; and plastic paste filler.

Small wing root fillets and most of those for fins and stabilizers, can best be developed using filler material; laid in by depositing a blob and squeegeeing it into place by running a finger along the joint (usually in several steps, with thorough drying in between).

RIGGING

Rigging is another of those daunting activities which deter the faint of heart from attacking vintage aircraft projects. One need only regard a 1913 Etrich Taube to appreciate the depth of the daunt.

RIGGING MATERIALS: The following materials are commonly used for static scale aircraft model rigging:

Wire: For aircraft, we are simulating wire, cabled or single-strand, which is often non-circular in cross-section as an aid to streamlining. Not surprisingly then, scale wire is a favored modelling material, be it steel, brass, or aluminum. But due to its stiffness, it is difficult to install in the taut condition, so that it in fact becomes what it represents: a dimensionally stable, structurally functional tension member.

Steel, for example, has an elastic modulus of 30×10^6 psi. Suppose that we use 10 mil diameter annealed mild steel wire, stretched to its elastic limit (30×10^3 psi) as a 5-inch long rigging wire on a 1/24th scale Parnall Platypus. The pull required is 2.3 lb. and the stretch is only 5 thou! Unless balanced by an equally tensioned wire in the opposite direction, you'll have a wing warper fer shure, once the model is released from the jig constraints. Ok, now suppose that it's really a 1/48th scale Vollrath Volksvogel: the proper scale wire is now 5 thou, so the pull to yield is down by 4 : 1/2 lb. The 5-inch length went to 2-1/2" though, so you only get 2-1/2 thou stretch to yield. The good news is that annealed steel wire can be pulled well beyond its elastic limit, allowing perhaps 25 mil stretch without any great increase in tension, or noticeable loss of cross-section (see a stress vs. strain diagram). So, annealed mild steel wire isn't too bad after all – stainless is even better, since it doesn't rust. In fact, steel wire is the preferred rigging material of many modellers.

Cable: Fine stainless steel cable, in two (or four) foot lengths down to 0.006" diameter (7 strands), is available in packets from Vintage Reproductions (The Emporium) in Colorado Springs.

Nylon Thread: The traditional model rigging material is of course thread: it's especially popular for smaller scales. But, typical cotton thread out of mom or spouse's sewing box won't do – it's furry and absorbs moisture.

What does work is nylon sewing thread, size A, available in black or grey. As it comes off the spool, the three strand thread (17 mil diameter) is appropriate for 1/48th and 1/32nd scale models. I (Alcorn) used it for my 1/32nd scale Rumpler CIV, whose rigging would still be taut from 1977, except that the top wing came off when I dropped it in 1985, and the movers finished the job in 1988. For smaller models, it can be unravelled and a single 5 mil strand used, as on George's 1/72nd scale Short Bomber.

Monofilament Nylon: Another good rigging material is monofilament nylon, which is usually transparent or smokey grey, and can be obtained in various sizes. Its primary disadvantage, relative to nylon thread, is that it is stiffer and slippery – and therefore more difficult to knot for anchoring.

Chapter VII: Assembly

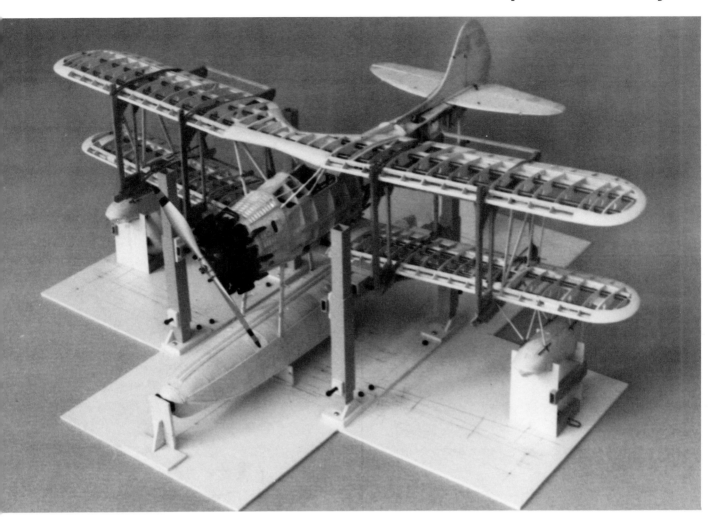

Here is George Lee's next *magnum opus* – a 1/16th scale Vought 03U-3 Corsair – fully jigged for component assembly and rigging. (George Lee photo)

I (Alcorn) used monofilament nylon for rigging my 1/16th scale Wedell-Williams #44. Since the actual wire was streamlined, I managed to find some Cortland Cobra flat (15 x 30 mil) fishing line which filled the bill (no one has yet offered me a satisfactory explanation for flat fishing line. Therefore, I presume that it's needed for catching flounder.) Smaller monofilament nylon, down to 5 mils or so, is ideal for smaller models.

Regardless of what material has been used, one of the saddest sights in all modeldom is an otherwise fine specimen whose rigging has gone limp. This usually signals its downhill decline to oblivion, since modellers are disinclined to redo old projects and few subsequent owners have the motivation, patience or skills to perform repairs. Wire (especially stainless) should be forever, so long as it was adequately tensioned (and anchored) originally, and the model itself hasn't distorted. Natural fibre thread is the worst offender, since it tends to creep with time and absorb moisture. Well stretched nylon has a good decades-long record: so long as it was pre-stretched before installation (by hanging weights from it for

Here the 1/32nd scale Rumpler, sans landing gear, is mounted upon its purpose-built rigging fixture. (Alcorn photo)

96   SCRATCH BUILT!

days beforehand). But, its hundred-year life is yet to be seen. (We should care? Of course we should, especially if our *magnum opus* is displayed in a museum, or perhaps destined to become a family heirloom.)

Untensioned Wire: One alternative: actually a cop-out, is to simulate rigging with stiff but untensioned wire, which is simply inserted into appropriate sockets: the wire is bent slightly to install, the sockets being deep enough to allow it to straighten when released. I (Alcorn) confess that I employed this disgraceful subterfuge for the wires of my 1/16th scale Laird Super Solution, now in the NASM. In this case, the model wings didn't require actual rigging, due to the firm top wing attachment to the fuselage and to the sturdy interplane struts.

Wires used in this manner must be of the work-hardened

Here is one last view of George's completed Keystone, again revealing the detail and workmanship which made this model the all-time top winner in any single IPMS/USA National event. (Ben Walker)

Chapter VII: Assembly 97

The rubber band secured balsa strips hold the wings gently but firmly against the interplane truss fixtures. (George Lee photo)

or heat-tempered variety, having a high yield stress. (The elastic modulus remains essentially unchanged by work-hardening, heat-treating, or alloying: E for steel remains about 30×10^6 psi, no matter what.) I (George) have found the best material for this application to be stainless steel orthodontic wire – that's right, the stuff they use for dental braces. It has a very high yield stress, comes in straight, precision-formed lengths, and can be obtained in a variety of diameters and flattened shapes. In particular, Unitek markets a line of 14-inch long "Permachrome" stainless orthodontic wire to the following specifications:

- Round Wire: Unitek No. 211-80 through 211-620: Diameter in mils (thou): 8 through 62.

- Flattened (Rectangular) Wire, with round edges: Unitek No. 241-820 through 241-536; size in mils (thou): 8 x 20 through 15 x 36, including six intermediate sizes.

This material, or at least ordering information, may be obtained from your friendly dentist.

For rigging larger models, such as my 1/32nd scale Keystone, I (George) have developed a technique for installing stiff orthodontic wire under some tension, using fine soft wire lashed to its ends, which have been notched. This is described in the subsection, Rigging Termination, below.

RIGGING JIGGING: In all but the simplest of jobs, the key to good rigging is good jigging. (Rigging in the jigging, sounds like…, well, never mind. Those nautical types are rude dudes. In any case, jigging in the rigging would be just as challenging.)

As discussed above for non-rigged aircraft, the first order of business is to prepare an alignment fixture for the fuselage and lower wings. The base is typically a board large enough to contain the entire A/C in plan view – and then some. As discussed earlier, the jigs for the fuselage – lower wings can be constructed using balsa. Means must be provided for holding the fuselage and lower wings snugly down against the jig register surfaces – even against the forthcoming rigging wire tensions. While rubber bands can be used, it may be best to lock things in place with glued-on top pieces. The accompanying photos of representative jigging setups are more eloquent than further verbiage.

For biplanes, further jigging is of course required to ensure correct position of the top wing, which is mounted upon the cabane and interplane struts. Usually, interplane jigging is important for establishing the proper gap and top wing incidence as the basis for final trimming and fitting of the cabane and interplane struts. Once the struts and top wing are set more or less in place, superstructure jigging is added to establish exact positioning of these elements. This jigging is usually larger and somewhat ad hoc, with the general appearance of scaffolding. Of course, it must permit access to the rigging for stringing and attachment. Again, grok at the accompanying photos for inspiration.

DEAD MAN ANCHOR: An aspect of rigging which must be anticipated during construction of the main aircraft components is provision of secure anchor points for the "wires." Thread allows the easiest solution. A double or triple knot can be tied in the end and stuffed through a very tight hole drilled in the host structure: the hole afterwards being filled with polystyrene paste or epoxy. For wire though, more elaborate procedures are required, for these reasons: since knots are not possible, it must be provided with some other type of "dead man;" the host anchorage must be strong to resist tension; and the wire must emerge from its anchorage pointing in the correct direction. A lump of solder makes a good dead man: alternatively, a hook at the end can be encased in a lump of epoxy. For such rigid dead men, it's best to insert the wire from the inside of the structure through a small hole. For

Arlo Schroeder is presenting a 1/32nd scale Grumman TBM-1C to President Bush, representing #46214 which the latter flew during his tour with VT-51 aboard the USS San Jacinto – including being shot down on 2 September 1944 following a raid on Chichi Jima in the Bonin Islands. On the right are Kathryn Schroeder and Tom Dietz of the NASM. Arlo adds that his Oval Office visit with the President was preceded by Michael Jackson and followed by Randy Travis and the lady who won the Iditrod dog sled race in Alaska. (White House photo)

SCRATCH BUILT!

NAKAJIMA A3N-1 (Type 90): This gem-like 1/48th scale A3N-1 is the handiwork of Ron Cole, one of the most promising of the younger scratch builders. He relates that the A3N-1 was a late 1930s trainer variant of the A2N-1 fighter. The fuselage shells were manufactured in the traditional manner. Due to thinness of the wings, Ron carved, filed and sanded them to final shape from rough epoxy castings. The leading edges were built up with primer to match the thickness of the wing ribs, which were simulated by strips of Chartpak drafting film. Multiple layers of Gunya Sanyo "Mister Primer" were then applied, and carefully sanded down to produce the interrib catenaries. At the 1992 Canadian IPMS National Convention in Ottawa, this model took First Place in 1/48th scale Scratch Built Aircraft – in a field of tough contenders.

monofilament nylon, I (Alcorn) had good luck with crimped ferrules as dead men. For my Wedell-Williams #44, anchor hard points were fabricated within the upper, forward fuselage structure. The ferrule anchored nylon line emerged through appropriate holes in the A/C skin and a cover patch was added over the anchorage. Needless to say, such embedded anchorages must all be made and the patches finish-painted, prior to rigging jigging.

PASS-THROUGH POINTS: In many cases, interplane rigging will zig-zag one or more times from bay to bay before its final "line" anchorage. With thread such a pass-through (change of direction) point at each bay can be achieved by simply wrapping the thread around the base of the interplane strut – so long as this yields the correct appearance: i.e., so long as the apex of the rigging angle occurs outside of the wing. Wires must pass through a hole, and as it is later drawn taut, it must be restrained from premature kinking by urging it through the hole using tweezers to overcome friction sticking.

RIGGING TERMINATION: For thread rigging, when no termination feature is required, either because of the model's scale or because the wire of the actual aircraft terminated in a hidden attachment, a hole can be drilled in the wing to receive the thread. When pulled taut, it can be fixed with epoxy or crazy glue introduced into the hole.

Rice: For the primary rigging of his monumental Il'ya Muromets, Bob Rice used 11 thou Trilene monofilament nylon fishing line, terminated in a most effective manner to represent turnbuckles. The end of each interplane line ("wire") was passed through a receiving hole in the strut end fitting, bent back and slipped through a short length (collar) of 1mm diameter brass tubing which had previously been fed onto the line. With the model "in the jigs", he cinched each line taut, added a drop of superglue into the collar and snipped off the tag end of the line. The strut ends were photoetched brass shapes (black anodized), bent to shape and slipped over the 1/32nd wire protruding from the ends of each laminated basswood strut. These details are well depicted in his March 1991 Fine Scale Modeler article.

Lee: For his Keystone, George Lee employed stiff, flat (8 x 20 thou) orthodontic stainless steel wire, installed under modest tension in the manner illustrated by the accompanying figure. The fine wire pigtails were passed through holes in buried wing hardpoints, and wrapped back around themselves (using tweezers). The square holes in the opposite side of the wing were later covered with scale inspection panels.

For rigging my Ninak, I tentatively (March 1993) settled upon the following approach for wire termination:

• 32 mil diameter brass tube, Hobby hanger item BT032 (15 mil I.D.) will serve as the body for turnbuckles and wire terminal fittings.

• clevis (fork) type fittings will be fashioned from 40 mil Hobby hanger brass tube (BT040): a slit is (Xacto) sawed into the tube end; the resulting forks are filed first on the outside; axle holes are pin vise-drilled near the end, perpendicular to the cut and flats; the tube end is cut off and dressed with a file and sand paper; and a wire shank is superglued in place. Tedious to the MAX for 50 or so clevi, but all part of the arcane pleasure of scratch building. Perhaps such clevi can be mass-produced on

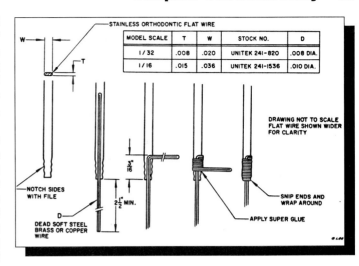

Termination of stiff wire for tension installation. (George Lee illustration)

a Unimat: better yet, maybe they're available from some commercial source as yet undiscovered by me.

• eyebolt anchor fittings will be either a single photoetched part, or a 12 mil diameter wire "eyebolt" from a (HO scale) Caboose Endgrab package (CA6504), supplied by Detail Associates, P.O. Box 5357, San Luis Obispo, CA 93403.

• a turnbuckle is then simulated with the appropriate length of BT032 tube, into one end of which is superglued an "eyebolt", the other end receiving a clevis shank, to be superglued during rigging installation;

• a rigging wire is terminated by supergluing into a short length of BT032 tube, the other end of which will receive a clevis fitting for later glueing;

• during rigging, a small wire "pin" is inserted through the clevis and its associated anchor point, and superglued in place;

• the wire, be it steel, brass or nylon, is tightened by pulling with tweezers at its terminaton, and the clevis shank is touched with superglue, anchoring it into the BT032 housing.

• an option to the clevis is simply to slip the wire through the BT032 tube, through the anchor eye, and loop it back through the tube. At installation, it is cinched to the appropriate tension, a drop of superglue placed in the tube and the wire snipped off. This is, of course, the Bob Rice approach.

CHAPTER VIII: PAINTING

AIR BRUSHING

Air brushing techniques for static scale model aircraft have been discussed in numerous "how to" books and articles. No two people have exactly the same technique nor the same perspectives on how to best describe their methods. However, some discussion is in order for completeness, so that the entire gamut of our procedures can be found within these pages. Since we presume that only experienced modellers would bother to purchase a book called Scratch Built!, we presuppose considerable knowledge of this crucial topic. As we've said before, we're often in awe of the finishing technique displayed on certain kit models seen at major IPMS functions. So, bear with us as we "preach to the choir."

NOTE: The discussion below under PAINT and SURFACE PREPARATION pertains to the techniques employed by Lee and Alcorn. The very effective "thin coat" approach favored by Peter Cooke is discussed above in Chapter V.

PAINT: While specialty enamel in pre-mixed colors is the traditional choice for most plastic modellers, we prefer acrylic lacquer – specifically, automotive finishes available from DuPont, Acme, Ditzler, R-M (Rinshed-Mason), Sherwin-Williams, etc. We find that lacquer is easier to work with, primarily on account of its fast drying characteristics. To some extent, our preference may be the legacy of having been weaned upon Testors and Comet "dope" during our impressionable solid (and flying) modelling days. (We even fondly recall the smell of the stuff!) A useful secondary advantage of lacquer is that enamel markings, weathering and other special effects can be superimposed upon the primary finish and removed at will with mineral spirits until the desired effect is obtained.

The principal drawback of automotive lacquer is that one must acquire a cabinet full of basic tinting colors, in pint containers or larger. Mixing to some exact color, especially camouflage hues, is something of an art in itself – often wasteful in terms of discarded failures, or excess quantity, once the precise color is achieved. Sometimes it is expedient to select a color from the dealer's catalog of automotive hues (actual color chips, arranged by auto manufacturer and model year) and let him mix it according to formula. This is, of course, rarely feasible for camouflage shades: RLM 70 Swartzgrun has not proven to be a big seller for passenger cars, even when fetchingly coordinated with 71 Dunkelgrun and 65 Hellblau.

Our only concession to jets, this Bill Bosworth model depicts a Heinkel He 162 "Volksjäger" of 3./JG 1 at War's end. The kill marks on the tail would indicate the pilot's score with other types, since no clear evidence exists that 162s ever downed an Allied aircraft. (Bosworth photo)

Chapter VIII: Painting 101

Although, like the "Volksjäger", the Dornier Do 335 "Pfiel" (Arrow) never had the chance to prove its worth in combat, it is a fascinating machine, with its tractor/pusher Jumo engine pair. (Bosworth photo)

EQUIPMENT: While not necessarily advocating it over other makes, we've always fancied the Paasche Model H airbrush, with the No. 1 tip and little open paint cup (yes, we've suffered many a spill). For compressors, I (Alcorn) use a Penncraft Model 6101 diaphragm unit. Some folks employ higher capacity pumps, in conjunction with a ballast tank and pressure regulator: others swear by bottle gas (dry air or C02). To be sure, bottles confer the advantages of mobility, quietness and dry, even gas supply. The February 1989 issue of *Fine Scale Modeller* contains an excellent article "Choosing An Airbrush Compressor", which covers 17 different makes/models.

We believe in adequate ventilation for all spraying activity and wearing of disposable filter masks for all but minor touch-up work.

SURFACE PREPARATION: This is, literally, the foundation of a fine finish, whether for automobile restoration or model building: in fact, the techniques are often similar, except for scale.

In Chapter IV, we discussed the frequent necessity for heavy block sanding of large vacuformed fuselage and wing shells, following integration of structure. Once the near final contours have been achieved, it is necessary to end up with smoothly sanded surfaces prior to detailing (panel scribing, etc.) and painting. Primer should be applied sparingly, not as a substitute for sanding, but to fill certain inevitable blemishes, including residual scratches, pits and joint anomalies. When sanding is complete, the primer should be thin, or even removed over much of the surface. No. 220 sandpaper is about the right grit at first, working down to #400 and #600 as visible surface scratches and other blemishes disappear.

Acrylic lacquer can be applied directly to bare styrene plastic, so long as it's not flooded on – bad practice in any event. We incline towards grey auto primer; a lacquer-based product well loaded with talc (or some such) filler to provide an easily sandable surface. Build it up gradually, to prevent "mud cracking", caused by surface melting of the plastic substrate in conjunction with surface tension within the drying paint cells. Incidentally, grey primer, brush applied "out of the can", is an excellent filler for deeper surface blemishes of various kinds.

AIR BRUSH SPRAYING: What can we say that you don't already know?

At your mother's knee, you learned that cleanliness is next to Godliness, especially as regards your airbrush: that it should be flushed with lacquer thinner after every use; and that the tip should be removed, disassembled and cleaned before putting away.

Also, from her you learned that more thin coats are better than a few thick ones. Later, from the kids in the alley, you also learned that this prevents the paint from destroying panel line crispness.

Then, from the girl next door you were taught to lay down dainty colors first, like Duck Egg Blue; with macho colors like Dark Earth and Middle Stone on top. You learned other things from her too, but they're outside the scope of this book.

MASKING: This is one of the make or break aspects of painting – a rough or uneven color division betrays the novice as surely as a visible shell seam.

Frisket paper is the standard of the world for air brush masking –unfortunately, some is as heavy as elephant hide. Our favorite brand has always been Redi-Frisket, which is thin and therefore conformable to double curved surfaces. But, even a good masking job doesn't guarantee a clean, crisp result. The key is to spray just enough paint to provide adequate color – but not enough to produce a noticeable ridge, once the frisket has been removed.

For the many camouflage schemes featuring diffuse and (usually) undulating color divisions, the raised mask technique can be employed. In this case, heavy paper (manila folder stock) is cut to the contour to be masked. One or more strips of masking tape is then laid along the down facing edge, set back about 1/8" on average. With the airbrush throttled well back and deftly handled so as to remain nearly perpendicular to the edge, the adjacent color is gently added. Some practice and experience is required to avoid the pitfalls of "flooding, shadow, and double images" beneath the overhanging mask. Generally, it's better to spray darker color over lighter. For those of sure hand and finely tuned air brush stroke, such color divisions can be made without the use of a mask.

Bob Rice has found that Chartpak crepe masking tape, 1/32nd inch wide, is ideal for masking thin lines, as on his Boeing 80A-1. The April 1988 *Fine Scale Modeler* contains an article on masking technique and materials.

NATURAL METAL FINISH: By this we usually mean bare "Alclad" aluminum skinning. As most of us have learned the hard way, this is one of the most difficult effects to realistically achieve with paint. Traditional aluminum paints yield a surface which is too dull and even. For this reason, certain specialty coatings have been marketed for the aircraft modeller which, if properly applied, can yield spectacularly effective results. Unfortunately, it's a limited market which is difficult for a small entrepreneur to exploit profitably, and all too few modellers have the knowledge and skills to realize the products' potential. Many of us recall Bob Moore, who developed, manufactured and marketed Liqua-Plate out of his home (or garage or some such): he even built and painted the impressive models which he used for his promo efforts. But, marketing factors eventually proved insurmountable.

At the time of this writing, commercially available products include the Model Master Metalizer lacquers from Testors, in both buffing and non-buffing colors; and the highly regarded Floquil silvers, which are also lacquer based. For further detail on effective application of these finishes, we refer you to the January 1991 *Fine Scale Modeler* article "Natural Metal Finishes."

WEATHERING AND SCALE EFFECT

Probably more has been written about weathering and other special surface effects than about any other aspect of modelling. We can't cover it all here: besides, we're not the experts, judging by some of the incredible effects seen not only on model aircraft, but on ships, trains, figures, armour and dioramas. One need look no further than Verlinden Publications' *Superdioramas*: gazing therein upon Lewis Pruneau's "Paris Gun" in its Kruppenwerke setting; or Bob Letterman's "Legacies" French village scene to realize that these guys are OUT OF CONTROL. In fact, we believe that the concept of weathering was first addressed by ship modellers, and later by the steam and whistle set. Its widespread application to aircraft is a relatively recent phenomenon.

WEATHERING: Either extreme of weathering can result in a distressing lack of realism to the trained eye. Typically, dedicated modellers strive to determine the exact colors of the prototype: indeed, it's one of the great preoccupations of the modelling crowd – ourselves included. So, having finally established these and having labored hundreds, perhaps thousands of hours over the creation itself, the normal emotional reaction is to produce a "clean" model, which shows off our skill and attention to detail "to the max." Having finally achieved this, to then tone the finish with greyed overspray, warp the fabric (either by surface contouring or tonal effects), scuff the wing walk area, splatter mud over the underside, chip the paint and exhaust stain the fuselage sides is tantamount to graffiti-ing the Sistine Chapel.

Our goal should be to produce a model whose photo, perhaps in color, is indistinguishable from one of the prototype. If it's a veteran DH-2 just back from a trip over the Somme, it will tax your skills to properly replicate its tatty appearance.

SCALE EFFECT: On the other hand, if you're portraying a factory fresh Heinkel He 111E, you have come face to face with "scale effect." The thing was decked out in a "splinter" scheme of RLM 61/62/63/65, whose hues are well portrayed in Merrick and Hitchcock's *tour de force*: *The Official Monogram Painting Guide to German Aircraft: 1935-1945*. But, as the authors explain in their Scale Effects section (pp. 6-7), you are well advised to reduce their intensity (chroma) by the addition of white – the smaller the scale, the greater the reduction. Furthermore, the markings –insignia, call letters, etc., – should be subtly greyed.

FADING: This is a major consideration for faithfully replicating veteran operational aircraft. The severity and nature of finish deterioration depends upon numerous factors, including color, quality of paint, harshness of the environment (heat, sunshine, rain, sand, saltwater, etc.), length of exposure and maintenance.

Modern acrylic lacquers and enamels are far more durable and colorfast than the finishes of yesteryear. We older folk recall all too well how the paint on prewar and early postwar automobiles faded and chalked. Reds and blues were the worst: it explains why most cars were either black, cream, pea green, tan or somber shades of livelier colors. And most cars were well treated, at least for the first few years, with washing, polishing and waxing being hallowed weekend rituals.

Chapter VIII: Painting   103

FOKKER T-2: This is Bill Bosworth's 1/32nd scale model of the famous aircraft which flew non-stop across the USA on 2/3 May 1923. Piloted by Lts. Oakley Kelly and John Macready, the journey from Roosevelt Field, Long Island to San Diego was made in 26 hours, 50 minutes. Donated to the Smithsonian by the War Department in 1924, it now hangs upstairs in the NASM. Aside from its imposing wing, one of the T-2's most engaging features is direct maintenance access to its Liberty engine from the cockpit. Incidentally, the T-2 was designed by Reinhold Platz of Fokker DrI and DVII fame. (Bosworth photo)

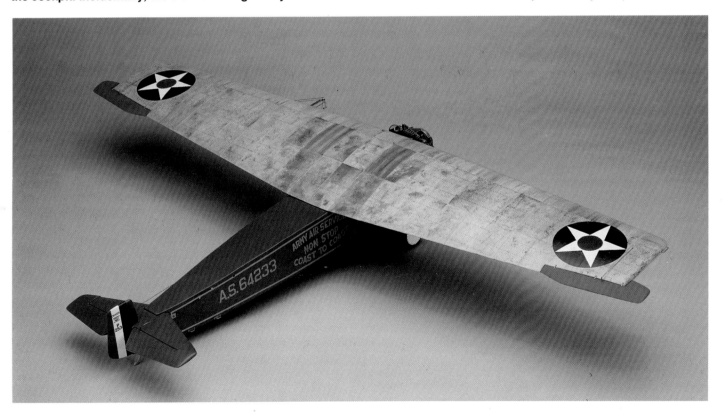

By contrast, camouflage finishes of World War II aircraft were continuously exposed to the elements and never cared for, except perhaps for the occasional washdown. Their non-specular finishes were notoriously prone to fading, scuffing and staining – little wonder that nowadays almost all flying vintage "warbirds" employ specular finishes as a concession to the practicalities of maintenance.

Dark Olive Drab No. 41 and the later AHA No. 613 used by the U.S. Army Air Force were among the worst "faders." While considerable allowance must be made for variation in film and subsequent fading of prints, color photos of veteran USAAF A/C reveal a most remarkable variety of color change. Most have faded considerably over their upper surfaces – often the hue seems to have drifted towards a reddish brown.

SCRATCH BUILT!

Ö. AVIATIK D.1 "BERG": George Lee's first scratch built plastic effort won Best of Show at the 1971 IPMS/USA National Convention in Atlanta. The wings were vacuformed sheet, ribs being glued directly to the top skin before insertion of brass spars. This 1/32nd scale model represents the 38th article of the second Osterreich Aviatik production order for the D.I: it was powered by a 200 hp Austro-Daimler engine. George's choice of an Austro-Hungarian subject was doubtless influenced by his friendship with Marty O'Conner. (Richard Hoyt)

Evidently, the extent and nature of fading was a strong function of the manufacturer and therefore of its specific chemical and pigment composition. At least two authentic examples of original wartime OD can be seen today at the Smithsonian's National Air and Space Museum in Washington, D.C.: the nose section of the Martin B-26 "Flak Bait" on display downtown; and their P-38J (L?) at Silver Hill (who knows, some day perhaps they'll break down and display "Flak Bait" in its entirety).

A remarkable example of olive drab fading is provided in Squadron/Signal's excellent B-17 Flying Fortress in Color by Steve Birdsall. In the color photo on page 24 showing the nose region of Schnozzle, the much faded factory paint contrasts dramatically with a freshly painted area where the new cheek gun window has been installed.

The modeller achieves the faded, as well as scale effect, by altering the "virgin" color, so meticulously prepared by matching to reference chips. The best approach is to prepare an ample quantity of the basic, but "scaled" color and then decant some off into one or more smaller bottles which are then "faded" by lightening and perhaps altering their hue. Then, the model is sprayed overall in the scaled color(s) and oversprayed as appropriate with the weathered tones.

Occasionally, a replacement panel should be indicated by masking off before the weathered overspray is applied. Alternatively, photos may indicate how repaired areas can be simulated by overspraying the faded surface with the initial (scale effect) hue – often in irregular "soft edged" splotches.

OTHER WEATHERED EFFECTS: Photos of service aircraft often reveal considerable evidence of localized staining, from oil leaks, engine exhaust, dirty hands, and machine gun blast. Such effects can be superimposed upon lacquer finishes with enamel paint, which can be removed with impunity until the desired effect is achieved.

Stains caused by engine oil being blown along fuselage (and nacelle) sides and undersides by the slipstream can be represented by brushing thinned-out dark grey in long streaks.

Spotty, splotchy stains can be produced by wet or dry brushing, as appropriate. For some effects, dark grey is appropriate, for others, darker tones of the basic color looks best. Sometimes, a subtle "handling discolorations" effect can be achieved by "nearly dry" brushing with lacquer thinner alone, to remove some of the weathered overspray.

Engine exhaust stains are usually most effectively represented by air brushing; various hues of well thinned russet/grey being applied with multiple, light fore and aft strokes. Sometimes a delicate greyish streak is laid down, over which reddish hues are lightly superimposed. It's perhaps worth reminding ourselves that some of the worst abuses of model finishing are created by overzealous treatment of exhaust stains – especially for liquid-cooled engines. While photos occasionally reveal a heavily blackened stain (resulting from carburetor problems or excessive oil consumption), the discoloration from well tuned engines results primarily from heat.

Chapter VIII: Painting 105

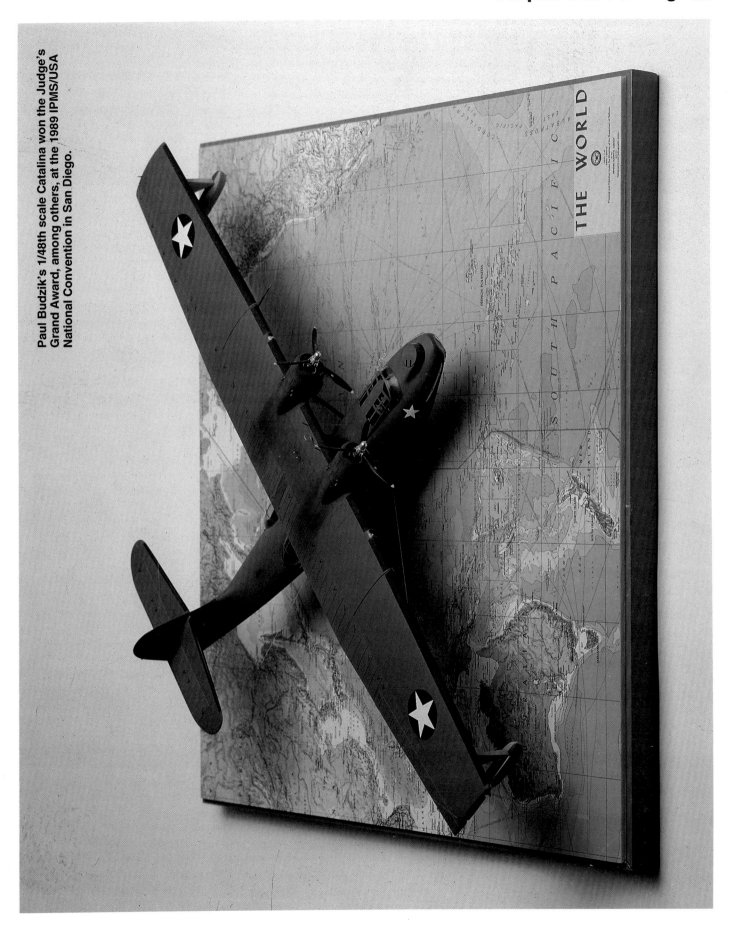

Paul Budzik's 1/48th scale Catalina won the Judge's Grand Award, among others, at the 1989 IPMS/USA National Convention in San Diego.

SCRATCH BUILT!

Clearly, Bill Bosworth is inclined towards the "last ditch stand" efforts of the Luftwaffe; this 1/32nd scale example representing a Ta 152H of JG 301, 1945. The Jumo 213E engined Ta 152H was the ultimate development of the efficacious and versatile Fw 190 series, developed under the direction of Dipl.Ing. Kurt Tank for high speed/altitude interception. (Bosworth photo)

Chapter VIII: Painting   107

The major components of this 1/32nd scale Bosworth model were vacuformed over hardwood forms. This "Nick" depicts a home defense night fighter of the 53rd Sentai. Pity that none of these handsome craft survive. (Bosworth photo)

SCRATCH BUILT!

VICKERS VIKING: One of the Brabazon Committee's choices for equipment of post-war British commercial airline service, the 36 place Viking was powered by British Hercules sleeve-valve engines: 161 were produced, primarily for BOAC. Alan Clark's 1/36th scale model is covered with 10 mil aluminum sheet, applied with contact cement! Double curved surfaces were formed after oven heating. (Alan Clark photo)

Representation of exposed metal from chipped or peeling paint is right up there with exhaust stains as the prime contributor to wretched excesses among overzealous weatherers. To be sure, you can find photographic evidence of certain aircraft which appear to be molting; but for most vets, evidence is limited to the occasional chip to base metal along wing leading edges, canopy frames, propeller blades and removable panels. Worn areas usually revealed evidence of the underlying primer, thinning down to base metal only in the most severe cases. Notable exceptions, however, include WWII Japanese aircraft which did sometimes reveal fairly extensive peeling and chipping, presumably from inadequate priming or inferior paint. Otherwise, dabs of aluminum paint should be applied sparingly and with thoughtful attention given to photo evidence. Rub pencil over the hardened aluminum paint and then buff lightly; otherwise, it just looks like aluminum paint!

GRAY OVERSPRAY: A very effective way to realistically deintensify the finish of a veteran aircraft is simply to lightly and evenly overspray the whole thing with a fine coat of well thinned dark gray – camouflage paint, markings – everything. Aside from overdoing it, the worst danger is having to later make spot repairs – it's tough to evenly restore the overspray.

SURFACE EFFECTS:

The final crucial maneuver for the model finish is surface treatment, to produce an appropriate "sheen"; from full specular to matte. Actually, no model should have either of these extremes: the shiniest, factory fresh NMF airliner should exhibit only a subdued specularity, while the weathered, chalky matte of a Memphis Belle should be tempered. Again, we're face to face with "scale effect." Matte is, after all, surface roughness, on the micron scale. When scaled to 1/32 or 1/48, we're back to "semi-gloss." As always, thoughtful reference to photos is your best guide – they have "scale effect" too.

As we have varied effects, so do we have various techniques for representing them. As a rule, we avoid using flat (matte) colors, such as are commercially available to the modeller, or can be produced in lacquer by adding flattening agent. Better control and uniformity of the final effect can be

Chapter VIII: Painting

FAIREY SWORDFISH: Alan Clarke's heavily weathered 1/24th scale model represents the anachronistic "Stringbag" – the Fleet Air Arm's instrument of victory at Narvik, Taranto and Cape Matapan, and over the Bismarck, many U-Boats and Axis vessels in the Mediterranean – yet also the ill-fated weapon whose vain assault against the Scharnhorst and Gneisenau during the February 1942 Channel Dash cost the lives of Lt.Cdr. Esmonde and many of his 827 Squadron comrades. (Alan Clark photo)

---

achieved by doing all of the color work in gloss and bringing out the final patina later.

An even, well controlled final coat of the basic (scaled) color is often good enough "as is." For example, the 1/32nd Keystone depicted herein was finished in this manner, since the paint of the original aircraft was not matte. The final coats were thin, though not overly thinned, yielding a surface which, though specular, was not really shiny. I (George) finish most of my models in this manner.

Quite an effective final patina can be achieved by lightly rubbing the painted surfaces overall with fine pumice – or "rough" aluminum oxide. The strokes should be fore and aft, especially on the wings and tail, to subtly simulate airstream scouring. This technique has the incidental advantage of removing any fine dust specs or spray powdering in the final coat. Peter Cooke prefers to simply "distress" the surface by rubbing with kitchen towelling, in the airstream direction.

Perhaps the most versatile, widely used technique for achieving the correct final patina is a light, thinned overspray of clear lacquer – or enamel. Here is the best time to introduce flattening agents – plus possibly that final tinge of deintensifying gray.

COLORS

HISTORICAL PERSPECTIVE: As all experienced aircraft modellers are also historical aviation enthusiasts, so are we reasonably knowledgeable about colors and markings – at least regarding eras/types which interest us. It wasn't always so, of course. It used to be that all WWII USAAF A/C were "olive drab and gray", British were "brown and green, with Duck Egg blue under sides", while German machines were almost universally "two shades of dark green above and light blue beneath." The Dark Ages. Antediluvian.

In all fairness though, the Harleyford folk tried to set us straight way back in the 1950s, with publication of *Aircraft Camouflage and Markings: 1907-1954*, by Bruce Robertson. While they supplied us with considerable markings information, their camouflage lore was – rudimentary. To quote from their chapter on Germany: "There was no general camouflage scheme in the Luftwaffe 1939-1945 . . . Dark green upper surfaces and light blue undersurfaces was typical of most German bombers in the early war years . . ." If that sounds vague, the color plate examples were downright ludicrous. An Me 109E "of the 1940-1941 period", for example, is depicted in purple uppers with brownish green blotches. Not surprisingly, they did somewhat better describing RAF colors, although the Dark Earth of the color plates is a bright light brown.

Lest you think that we're picking on Harleyford, we'll cite one other glaring example. Turn to the Heraldry and Camouflage section of William Green's, *Warplanes of the Third Reich* of 1970 – an epic, scholarly tome if ever there was one. Now turn to the third color plate (p.19) covering Bf-109E's. Hellblau 65 has become a deep, lurid shade of indigo! Except for desert sand examples, almost all of the upper surfaces were depicted in tones of green – perhaps in deference to the author's name. No 70/02/65 or 74/75/76 schemes in sight.

To be sure, considering Harleyford, we were in the uncritical, pre-research era as regards color; however, since many authors and modellers of that era had seen the real machines, some at least must have had a more refined sense of camouflage colors than the plates suggest. The worst problem was evidently that of faithfully reproducing a subtle hue in printers ink: doubtless exacerbated by poor communication between author and publisher; if "dark brown" or "dark green" was the printers instruction, he simply used what he had readily available.

WWII GERMAN COLORS: And then, along came Geoffrey Pentland, Richard Smith, John Gallaspy, Ken Merrick and Thomas Hitchcock. Smith, Gallaspy and Merrick's three-volume *Luftwaffe Colors* from Kookaburra was a landmark, while *The Official Monogram Painting Guide To German Aircraft: 1935-1945* by Merrick and Hitchcock is surely the definitive reference. Within the heavy loose leaf pages of this wondrous tome lie just about every nuance of German camouflage we'll ever need to know. Presented chronologically by type, the material is beautifully presented with quality photo examples, both black and white and color; RLM painting guides for each major A/C type; references to RLM orders; quality color profiles; and most important of all, pasted-in color chips made from actual paint. The dedication of these fellows over the years has been a source of admiration, awe and appreciation by those of us who share their interest in such arcane matters.

In their earlier volumes, we occasionally felt that they were reaching a bit to state categorically that some A/C shown in a fuzzy 4th generation black and white photo was painted in such and such colors.

But, as their scholarship advanced over the years, our skepticism gave way to great confidence in *The Official Monogram Painting Guide*. When they're not sure even now, they're candid enough to admit it. We are indeed grateful to them.

It was these fellows who systematically developed the thesis that many Luftwaffe fighters of the 1940-1941 era had had their 71 Dunkelgrun upper portions replaced by the lighter 02 RLM Gray – eventually substantiated by surviving fragment examples and RLM directives. Likewise, it was they who cleared a great deal of confusion and misrepresentation by elucidating the definitive 74/75/76 "gray" scheme for Luftwaffe fighters from 1941.

USAAC/AAF COLORS: It was Dana Bell who put us right on USAAC/AAF colors and markings, with the publication of his two volume *Air Force Colors* (Squadron/Signal). Volume One covers the Air Corps and Air Force through 1941, while Volume Two covers the ETO and MTO through 1945. One can only hope that someday the series will be completed with Volumes on the Pacific. While actual paint chips are not included, he provides a comprehensive discussion and listing of colors, keyed to their Federal Standard 595a equivalents. Despite the titles, his volumes go well beyond simply describing color use: his photo/text coverage of markings is thorough and well researched. This includes theatre directives and practice, national insignia changes; and extensive coverage of group/squadron markings.

US NAVY COLORS: Dana Bell teamed with Berkley Jackson and William Riley to produce *Navy Air Colors: United States Navy, Marine Corps, and Coast Guard Aircraft Camouflage and Markings*, Volume 1: 1911-1945, also by Squadron/Signal Publications. While no color chips are provided within, color identities are keyed to Federal Standard 595a.

More recently (1987), Monogram Aviation Publications has come forth with their multivolume series *The Official Monogram US Navy and Marine Corps Aircraft Color Guide*. Volume 1 covers the 1911-39 period; Volume 2, 1940-49; Volume 3, 1950-59; and Volume 4, 1960 to present. Multichapter sections of these scholarly, hard cover tomes are devoted to aircraft painting; national insignia; and markings for identification/recognition, maintenance/safety, and meritorious achievement. Some color photos and illustrations are included, plus a valuable page of color chips.

Both the Squadron/Signal and Monogram Publications feature excellent selections of aircraft photographs – reason enough to purchase these books. We prefer the Squadron/Signal color aircraft illustrations, but the Monogram color chips are a great convenience, and their coverage of official directives is of considerable interest to cognoscenti.

WWII RAF COLORS: A little booklet published by Arms and Armour Press for the RAF Museum entitled *British Aviation Colours of World War Two* includes a sheet of excellent paint chips.

WWI COLORS: In Windsock and Datafiles, Ray Rimell has distilled the essence of modern research on the subject of colors for WWI aircraft; citing Methuen equivalents, and often providing paint chips keyed to period documentation and surviving samples.

In its spring 1968 journal (Vol. 9, No. 1), *Cross & Cockade* published Part I of "Project Butterfly": the Standard French Camouflage System of 1918, prepared by A.D. Toelle, H.D. Hastings and Bergen Hardesty. Part II which appeared in the Summer 1972 journal (Vol. 13, No. 2) added much additional material, all of great interest and value to the dedicated enthusiast/modeller.

WWII JAPANESE COLORS/MARKINGS: Koku-Fan's *Japanese Imperial Army and Navy Aircraft Color, Marking* (illustrated No. 42) is an excellent compendium of this fascinating subject. Many color plates are provided, supplemented with a fine selection of photographs. However, no paint chips are included, and if reference is made to Munsell, Methuen

Chapter VIII: Painting 111

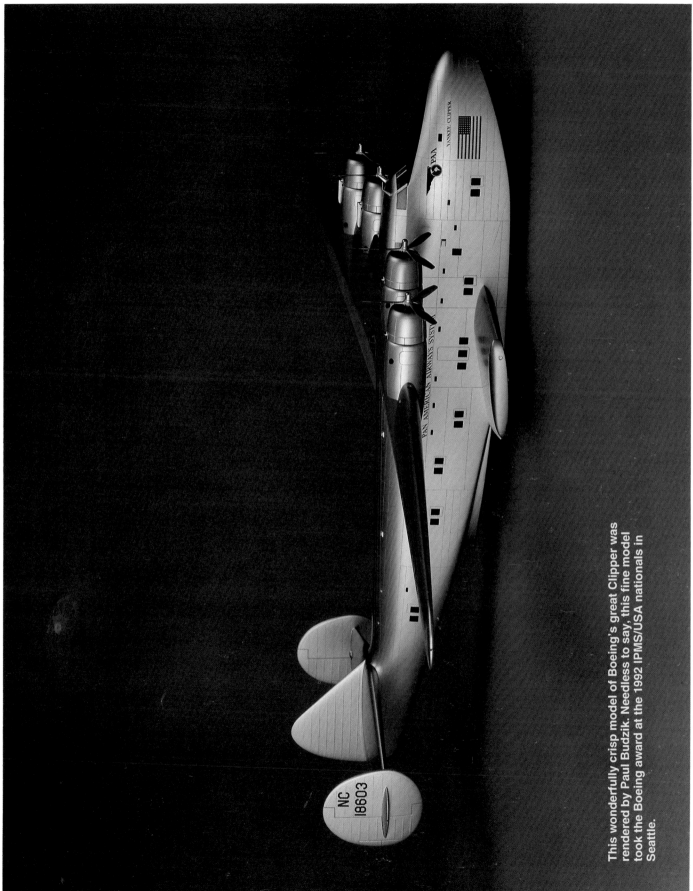

This wonderfully crisp model of Boeing's great Clipper was rendered by Paul Budzik. Needless to say, this fine model took the Boeing award at the 1992 IPMS/USA nationals in Seattle.

other color standards it is not revealed to those of us who don't read Japanese.

Another source, now long out of print, is Aero Publishers pair by Donald Thorpe: *Japanese Army Air Force Camouflage and Markings, WWII* (1968) and "*. . . Naval Air Force* (1977) . . ." Interior colors were addressed by Monogram's close-up 14 and 15: *Japanese Cockpit Interiors*, Part I (1976) and Part II (1977), produced by Robert Mikesh.

For those who failed to obtain the above titles, take heart: Robert Mikesh is now (1992) preparing a color guide for Japanese aircraft which will contain sample paint chips.

Incidentally, a fascinating and useful essay entitled Japanese for Modelers by LCDR. William F. Hardin was included as a ten part feature in IPMS Update, beginning with Vol. 10, No. 5 of May 1975.

COLOR STANDARD REFERENCE SOURCES: To obtain copies of certain color standard references cited herein, Robert Mikesh has kindly provided the following addresses:

Munsell Color Company, Inc.
Division of Kollmorgen Corp.
2441 North Calvert Street
Baltimore, MD 21218

Their complete color guide, suitable for institutional use, is rather expensive. However, the Munsell 11 Chart Student Set is available for around $31 (March 92). It should be obtained with value, and chroma notation on the back.

Federal Standard Colors (FS595B)
General Services Administration
Specifications Activity
7th and D Street, S.W.
Room 6654
Washington, DC 20406

A nice set of color chips, in elongated "fan" format, is available as Item No. 7690-01-162-2210.

Pantone, Inc.
55 Knickerbocker Road
Moonachie, NJ 07074

This organization can provide a handy booklet with a series of color shades. These tear off in rectangles about 1/2 x 3/4 inches, about ten chips per color, for comparison with existing samples.

Methuen Handbook of Colour, by
A. Kornerup and J. H. Wanscher
Methuen and Co., Ltd.

This book, containing over 1260 color samples, is currently (1991) available at around £32 from:

Peter Grose Ltd.
P.O. Box 18
Mayhill, Monmouth,
Gwent NP54YD
Wales, United Kingdom

WEDELL-WILLIAMS #44: I (Alcorn) built a 1/16th scale model of this legendary air racer for the NASM, in its 1933 Thompson trophy winning guise (Wasp Jr.). While I had obtained many fine black and white photos of the machine, my knowledge of its 1933-1934 red colors was limited to a few vague descriptions in the literature, plus the opinions of a few air race enthusiasts. Figuring that I'd done all I could in the name of authenticity, I proceeded to paint it.

Shortly thereafter, but prior to application of markings, rigging and final details, I took it to an AAHS function in Los Angeles (circa 1979). While I was standing around the display table nursing a Michelob, some surly old peg-leg came up and inquired, "That yours?" I proudly replied in the affirmative, whereupon he volunteered, "Well, the color's wrong!" Somewhat taken aback and slightly annoyed at his dogmatic judgement, I asked how he knew. "I saw it", was his curt reply. "Oh, right" I thought, here's this guy who may or may not have glimpsed it 45 years ago and now pronounces my red as "too bright." The world is full of such pompous know-it-alls. Well, his name was Rudy Profant; he had indeed watched it crash at the 1934 Thompson (killing Doug Davis); had stormed the smoldering wreckage along with a hundred other bloodthirsty urchins; and had made off with a piece of red fabric which he'd kept ever since in a book. Hmm . . . "Uhh, pleased to meet you, Rudy." Well, I couldn't get him to tear off just a little corner, or loan the piece to me (would they send you the Shroud of Turin just to look at? Well maybe now they would). But, he did send me a color chip for a 1953 Buick which he said matched perfectly. Needless to say, it's now painted Duco 2259H: Seminole Red.

CHAPTER IX: MARKINGS

SILK SCREENING

If casting crystal clear canopies, machining minute Maxims or winding wimpy wire wheels has ceased to challenge you, silk screening your own multicolor decals may get the old juices flowing again. If this doesn't drive you into the Twilight Zone, you can't be driven. All you need really are rock steady hands, hard wired nerves, the Patience of Job, the Stolidity of the Sphinx, aerospace quality tooling, a Swiss bank account and the Luck O' the Irish. But, face it, if Microscale hasn't yet seen fit to produce a 1/32nd sheet for Costes and Bellontes' 1930 Breguet Super Bidon, what choice have you?

The technique which we use involves the following basic steps: art work preparation; photo reduction; silk screen preparation; overlay screening; overspraying and application of decals. The crucial process is color overlay screening.

ART WORK: The first step is to draw the marking – first in pencil, then in India ink – on stiff, quality paper, usually to a larger scale than that of the finished product. The smaller and more tedious the marking, the greater the artwork magnification: x 4 or x 8 is common, especially for "tinywritten." For a single color marking, such as black or white stencilling on the real thing, you just reproduce the designs, remembering that the final decal will be no better than your original. So, care and precision is essential: use straight edges, French curves, circle templates, compasses – whatever is necessary to produce an accurate, crisp facsimile. If you make a mistake in ink, it's o.k. to use white Liquid Paper Correction Fluid – only your pride will suffer. Put several markings on a single sheet, even if some will later be screened red, some white and some black. Later, you can screen sheets of each color and use only the appropriate ones.

If color overlay is required, then you must produce separate, but precisely indexed art for each color. Suppose, for example, that you are preparing markings for a 1/32nd scale Boulton-Paul Defiant (using MAP PP3014, natch) of No. 264

WEDELL-WILLIAMS #44 (NASM photo)

SCRATCH BUILT!

ARMSTRONG-WHITWORTH SISKIN Mk.IIIA: This RAF stalwart was rendered in 1970 by Harry Woodman, a pioneer in the use of "plasticard" for scratch-built modelling. Comprehensive details of the model's manufacture are given in the May 1971 issue of Scale Models; including 1/48th scale drawings, and patterns for the fuselage structure and skin. (Ron Moulton photo)

Squadron, as used during those fateful, fatal days of August 1940 when operating from Hornchurch/Manston. You'll be making the national insignia for the fuselage, upper wings and fin; the large squadron (PS) and individual A/C fuselage letters; the A/C serial number; and certain tinywritten such as the W/T stencils. (For the sake of argument, let's assume that no satisfactory national insignia was available from commercial decal sheets, kits included.)

For the first sheet, to x 2 scale, lay out, in pencil, all the large "solid" color elements: circles for the fuselage national insignia (to be yellow); for the upper wing roundel (to be blue); for the fins (to be white); and for the fuselage letters (to be grey). Outline the fuselage circle (and large letters in ink) and fill them in with a brush. Also, make two index marks (small crosses) at diagonal corners of the artwork field.

Now, working on a light table, place a second sheet of sturdy inking paper over the first and securely tape it down. Carefully pick out the center-points for the circles and index marks, and with pencil lead and compass, scribe circles on the top sheet.

Now, remove the second sheet from the light table, lay out (with ink compass) and fill in the white region for the fuselage insignia. This should be slightly undersize for the blue region (say, 1/16" on the radius, at x 2 scale) and completely filled in (no need to leave a "hole" for the later red dot). In this manner, when the blue is later silk screened over the white, no white will peek through its outer edge. Also, trace, lay out and fill in the entire (white) region of the fin flash. Remember to ink in the two index marks.

Now, tape a third sheet over the first on the light table, repeating the inking process for the areas to be blue: the ring for the fuselage insignia; the large circle for the upper wing roundel; and the aft portion of the fin flashes. Finally, prepare a fourth x 2 overlay sheet for the red areas: the fuselage and wing meatballs; and the forward regions of the fin flashes.

An independent x 4 artwork sheet should be prepared for

Chapter IX: Markings 115

CURTISS A-12 SHRIKE: This Bob Davies model represents an A12 in use by the 90th Squadron, 3rd Attack Group, in the mid-1930s. (Partain photo)

the smaller markings (A/C serial number, W/T marks, etc.), unless you are skilled enough to crisply produce these on the x 2 sheet.

PHOTO REDUCTION: Next, take the inked artwork to a reputable photo shop, to obtain 1:1 scale negative transparencies of each. Remember to erase all pencil marks on the originals.

SILK SCREENS: Some hard cases have been known to prepare their own silk screens. But we "normal" folk employ the services of a professional silk screen shop, who simply "burn" the images through the emulsion using our photo transparencies. Caution them to place the successive overlays on the same location for each screen – this will make your screening work somewhat easier.

SCREENING: We suppose that the main reason silk screening looms as such a challenge to us scratch builders is that we engage in it so infrequently. Each time, we recall the technique only dimly and have misplaced half of our equipment – use it, or lose it, so they say. The upside though is that we've also half forgotten the frustration, the waste, the mess – and our resolve: "never again!"

In principle however, screening decals is simple: like so many activities, from fixing the faucet to launching a satellite, the difficulties lie in the details of implementation. For a single color screen, all you need is a means of holding the paper firmly down; a means of holding the screen over, but just above the paper; the framed screen itself; a blob of appropriate paint; and a rubber squeegee. As you drag the paint over the screen design with the squeegee, it is forced down through the open mesh onto the paper. After the squeegee has passed, the taut screen lifts off the paper, leaving behind the finely detailed image.

Yeah, well . . . in practice, the paint is too thin; the mesh is partially clogged; the screen is poorly indexed, slips and doesn't lift off the paper. The result is a smeared image, paint all over the place, and you in a rude mood. When the statistical chances for success are exponentially diminished by some power of the number of colors to be applied, your chances for success are comparable to those in the Lottery.

The Rig: As with vacuforming, the first prerequisite for success is proper equipment: the screening apparatus. A proper rig consists of a vacuum box for holding down the paper and a hinged clamp for holding the wooden silk screen frame. While such equipment can be custom made, commercially manufactured rigs are available from large specialty art supply dealers.

Paper: For our purposes, uncoated decal paper is required. It's simply blank paper coated with a layer of adhesive – the same slimy stuff found on the bottom of any wet transfer decal.

Color Overlay: For color overlay screening, registration is the crucial aspect: assuring that each succeeding color pattern lies directly over the one below. Each piece of decal paper must have been cut square (rectangular), and is placed with two orthogonal edges against the raised "carpenters square" ridges on the top of the screening box. Only your own bitter experience will tell you how many extra first color sheets must be made for each number of color overlays. A typical scenario though might be: x 2 for one color overlay; x 4 for two; x 8 for three, etc. The actual number required will depend upon your

SCRATCH BUILT!

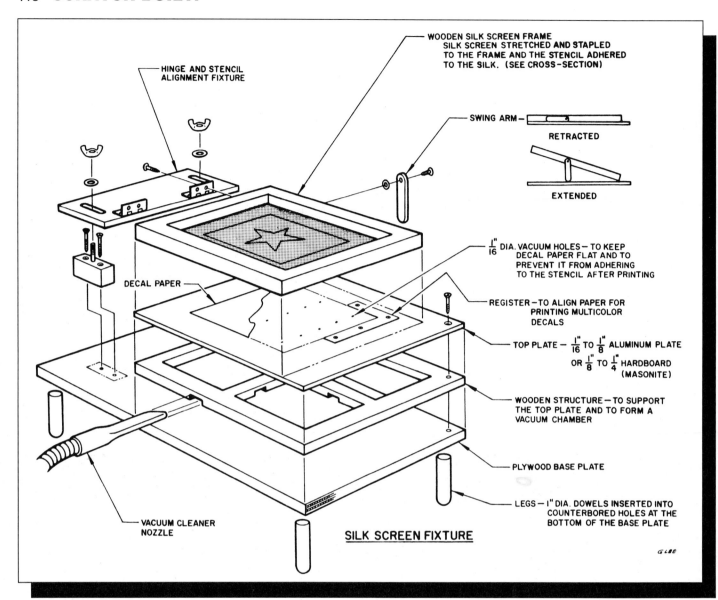

(George Lee illustration)

skill and luck, plus size, number and complexity of the pattern elements. In any case, decal paper and paint are cheap: the time and frustration of running short on multicolor overlays is not.

A decal sheet with the first color applied is left on the fixture, snug against the edge locators. The screen for the second color (first overlay – yellow for the Defiant examples) is located over this sheet, using the cross index marks, and clamped in place. Now, it can be rotated up and laid back down on this sheet, or subsequent ones, and still register perfectly – if the hinge/clamp has no slop.

Technique: As with most techniques, instruction from an expert and hands on experience are better than written descriptions. Nevertheless, we'll try . . .

Normally, silk screen paint is the proper consistency as it comes out of the jar or can – thick and even textured.

Place a large blob on the screen, on an area where no design has been etched. With the screen down over the paper, simply drag the squeegee firmly and smoothly across the design, with the paint ahead of it – the squeegee should make an angle of about 70-75 degrees with the screen, with the paint "in the angle." The pressure forces the screen down against the paper and the paint through the mesh. The combination of pressure and paint thickness should prevent paint from getting under the design edges, so that a crisp image results and no cleaning of the screen underside is necessary before the next application. Incidentally, never make more than one draw across any given sheet.

If for some reason – often mysterious – some paint has gotten beneath the screen mesh, it is wiped off with a rag which has been dampened with mineral spirits. Otherwise, just keep going until you've produced the requisite number of sheets of that color.

Cleaning the screen upon completion of a color is messy

Chapter IX: Markings

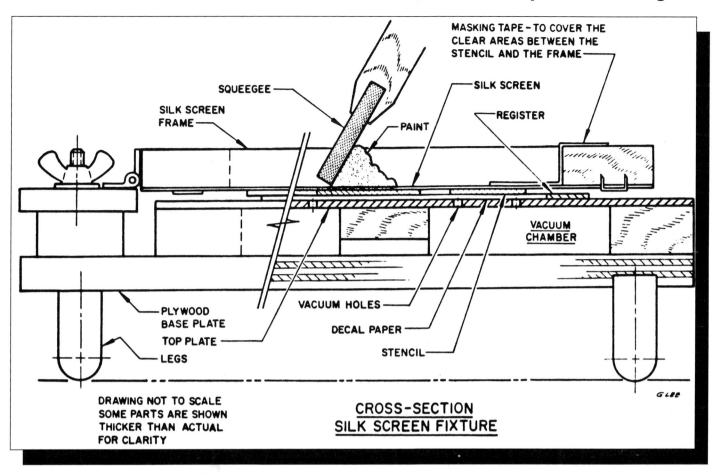

(George Lee illustration)

at best. The excess paint is first scraped off with the squeegee, followed by cleaning with a solvent-wetted rag. Needless to say, it is imperative that you clean all of the pores of the mesh.

DECAL COMPLETION: Once acceptable markings have been screened upon the paper, they must be oversprayed with clear lacquer to provide transfer strength. In fact, if the design is separate elements without background color, such as the fuselage serial of our Defiant example, then the clear lacquer skin is necessary to hold it all together during transfer. Once the lacquer has dried, the design can be cut out of the field with an Xacto knife and wet transfer applied just like any other decal.

ALTERNATIVES FOR MULTICOLOR DECALS: Since your chances of producing a good decal are inversely proportional to the number of color overlays, it's sometimes worthwhile to consider certain low risk options.

One is to prepare two-part decals: one applied atop the other on the model surface. A simple example would be the fuselage insignia of our Defiant: you could apply the yellow/blue/white portion and later add the red dot as a separate decal.

Another possibility for our Defiant might be to mask and air brush the yellow background circle directly onto the fuselage, later applying the blue/white/red decal – possibly in two parts as noted above.

AIR BRUSHED MARKINGS

Often it is expedient to simply air brush markings directly onto the painted surface of the model, having suitably masked with frisket paper. Appropriate features may include national insignia, large letters/numerals and color bands. The advantage is two-fold: you avoid the tedious silk screen process and the finishing techniques required to "blend" any decal into the background finish. Air brushing is usually easy, so masking is the limiting factor.

MASKING: First, the outline of the feature to be produced must be laid out on the painted model surface with a pencil: HB lead is usually best. Use a compass for circles – a tiny hole in the center is a useful benchmark, and can later be covered.

Although tedious, the best way to mark (or scribe) a line around the cross-section of an oval fuselage (such as for the white band of a mid-war Spitfire) is to make a basswood or plastic template, filleted with backing gussets so it rests perpendicular to the fuselage centerline. After pencil marking, masking can then be performed by first laying down thin frisket strip, stretched as necessary to conform to the double curved color edge.

Features with straight edges, such as a large letter "A" or numeral "4" can be masked with ruler-cut frisket, laid down in overlapping strips.

Chapter IX: Markings

For certain shapes, such as an insignia star, the pattern can be drawn on the frisket beforehand; the frisket carefully applied to ensure position and alignment; and the pattern then carefully cut out with gentle strokes of a sharp #11 blade, guided by a flexible straight edge (plastic, if the curve is considerable). An alternative is to pencil lay out the star (within a guide circle) after the frisket is applied.

PAINTING: Air brushing the color is, of itself, a trivial operation. We must simply remember to lay down only enough paint to develop its full intensity: too much and we run the risk of leaving excessive edge ridges when the frisket is removed.

FINISHING: Despite our best efforts, some ridge may be in evidence at the margin of the new color. The usual way to remove it is with a small piece of crocus cloth or #600 wet-or-dry sandpaper, lightly applied so as to just level the rim. Thus, when the final overall clear coat is applied, no surface contour will be visible at the color division.

LAIRD "SUPER SOLUTION": Matty Laird's hurriedly-built "Solution" was completed just in time to win the 1930 Thompson Trophy, in the capable hands of Charles "Speed" Holman. For next year's National Air Races, Laird carried the fixed gear, Wasp Jr.-powered, biplane concept forward in the "Solution's" successor. Jimmy Doolittle's flying and navigational skills guided the little craft from Burbank to Cleveland (with pit stops at Albuquerque and Kansas City) in 9 hours, 10 minutes, to win the first Bendix Trophy race. Sensing a further opportunity, he refuelled and continued on to Newark, setting a transcontinental record of 11 hours, 16 minutes at an average speed of 217 mph. He then promptly hopped back in and returned to Cleveland for the festivities. This 1/16th scale model was built by John Alcorn for the NASM's 1976 opening. (NASM photo)

Opposite:
Boeing 80A-1 Details. (Rice photo)

SCRATCH BUILT!

HAND PAINTING

Despite all the marvels of commercial decals, silk screening and air brushing, there is still a need for a certain amount of hand painting. In fact, some tradition-steeped folk still do much of their markings by hand – national insignia and all. They can be every bit as satisfactory as the results by other methods – except of course for really small lettering. I (Alcorn) have a wooden (solid) 1/24th scale Hawker Hurricane whose national insignia and squadron letters were done by hand – with lacquer. It looks fine – but the markings took an eternity!

Aside from weathering effects (discussed elsewhere) hand painting remains the best means of adding detail and highlight color to certain intricate, non-geometric designs, such as "nose art." On my 1/32nd scale A-20A, I (Alcorn) did most of "Daisy Mae" by hand, using enamel, applied with a No. 00 sable brush.

Speaking of hand painting: how many of us own a gen-u-ine George Lee tie? (or, can tie a decent Windsor knot, for that matter). As if he weren't busy enough, George drew and silk-screened family Christmas cards every season for many years, usually with a humorous theme.

EPILOGUE

Who can say where our hobby will go from here? Materials, techniques; subject matter all drift inexorably towards ever increasing sophistication. After each IPMS/USA national contest, for example, the participants return home resolved to go a step beyond for the next. A tad more detail, a further touch of realism and a whammo subject. Let's hope it never ends.

But, evolve it must. Subject matter is a prime driver. Our hobby began with WWI subjects and has been dominated by the aircraft of WWII for 45 years. But, modern jets have taken over, as evidenced by the offerings of plastic kit manufacturers. Surely spacecraft, both real and imaginary, will dominate the tables of future contests. Oh sure, some contrarian scratch builder will always reach back in time for some ancient subject which has captured his imagination. The day will even come when a P-51D will be a radical, attention-grabbing entry. "Wow, dig that clumsy windmill! And that stupid stick they used to control it. Those analog instruments are a riot! And, those awkward lines: no wonder it could barely do Mach 0.6."

(I wouldn't even hazard a stab at the lingo of Earth date 5427. Let's just hope that they still know about Mustangs and perhaps still have an example or two to snicker at.)

We've graduated from wood to high impact polystyrene for primary components. While the latter material seems ideal, perhaps a new generation of plastics will emerge. But, techniques seem to offer far greater potential for change, both for commercial manufacture and handicraft. We've witnessed the appearance of photoetched parts, decals of exquisite quality, injection molded kits of amazing complexity, and the resin cast marvels of Peter Cooke. Doubtless, new finishing materials and techniques will emerge to render our present air brushing efforts obsolete. Rub on finishes, for example, would be convenient, easy to apply, unmessy, and amenable to various tonal effects.

Meanwhile, we've had a grand time doing our thing in the old fashioned way, but always pushing the state-of-the-art to the limits of our skill and imagination.

1977 IPMS/USA National Convention, San Fransisco: Arlo Schroeder is presenting the Best of Show trophy to Larry Templeton, for his R5D maintenance diorama. 1977 Convention Chairman George Lee is on the left: Lloyd Jones and John Alcorn on the right. (IPMS photo)

POSTSCRIPT: MODEL PORTRAIT PHOTOGRAPHY

Having come to grips with this subject during preparation of Scratch Built!, I (Alcorn) am motivated to volunteer a few observations (prejudices).

While optically adequate equipment is certainly a prerequisite for good model photography, fancier is not necessarily better. In fact, in my reactionary view, electronic gimcrackery is tantamount to "not under operator control," at least for static studio subjects. If you know the basics of photography, you understand focal length, shutter speed, film speed, depth of field, lens opening ("f" stop), focus, and their interrelationships. You don't need an on-board microprocessor to sort it all out; and certainly not a built-in dynamo to rewind the film!

Studio cameras are great, since we all recognize that the larger the image (negative), the greater the detail which can be captured. But, for up to 8 x 10 size prints, a 35 mm transparency or negative can be adequate, so long as moderate speed film is used, and other basics are not ignored. The Peter Cooke and Bob Rice photos in this book were all taken with a 35 mm camera. Tripod mounting is essential, both for composition and image sharpness.

For close-up subjects such as model portraits, and especially model detail shots, a single lens reflex (SLR) camera is almost a necessity: also, focus distance must be compatible with the subject proximity required. So much for equipment.

First, we must ask ourselves: "what am I trying to accomplish?" Generally, we want a straightforward shot which shows the model in an attractive, natural pose, revealing its salient attributes in crisp, clear detail. Dramatic poses featuring backlighting, fish-eye lenses, defocus and hyper-perspective are best left for advertising and photo magazines/exhibits.

Equipment aside, the first and most often violated prerequisite is adequate lighting of the subject. This allows the maximum potential for depth of field, at a given film speed. But mainly, it gives contrast and vividness to the subject. Harsh shadows should usually be avoided, since they are distracting and leave certain portions of the model in the dark. Therefore, multiple source lighting is advised: take special care that important and prominent details, such as radial engines and markings, are well lighted. At the other extreme however, beware of lighting so uniform as to lose the sense of three-dimensionality provided by light/dark contrast and highlights.

One of my pet peeves, for actual aircraft as well as models, is lack of perspective through use of a telephoto lens. Indeed, its use is now so all-pervasive that aviation artists often unconsciously replicate this very unnatural appearance. Telephoto has its place, of course, for photo documentation as a basis for plans preparation; and for capturing distant in-flight subjects, as at airshows.

The pose (viewpoint) selected should be dictated by emphasis of a model's virtues as well as by esthetics. As a rule, it's wise to photograph from several angles and at varying light settings, so that a selection can later be made from prints.

Perhaps the most universally satisfying model portraits are those taken simultaneously from above, in front and to one side: in this manner the subject is revealed in an attractive manner which tends to fill a rectangular space. Linear shots should normally be avoided, such as (almost) side views or those looking edge-on at the wing(s). There are exceptions to be sure, such as the attractive shots of Peter Cooke's "Johnnie Johnson" and 21-T Spitfire MkIXs shown in Chapter V. Also, his evocative "eye level" portraits of the Royal Navy Sea Fury (#10) and Mosquitos are very effective. But, it's hard to beat those lovely poses of Rice's Il'ya and Bellanca "Cruisair"; Bosworth's Sikorsky S39B, He162, "Volksjäger" and Dornier 335; and George's Keystone featured within these pages.

As a general rule for model portraits, the background should be of a neutral, contrasting color; have a minimum of texture, and not exhibit "distractions." The neutral, textureless color is usually best achieved by a smooth cloth, draped over a raised object behind the model, so as to provide a featureless background. "Eighteen percent grey" is favored in the photo biz, though other colors may better suit a particular model.

Now, if you revisit the model portraits within these pages, you'll discover that we have violated each of the above canons on occasion. Our rejoinder is yet another canon: "Consistency is the hobgoblin of little minds." Besides, we coauthors did not have total control over the photos-even of our own models! For example, many are now in museums and hence unavailable for all practical purposes. Such photos as were hastily taken before delivery are all that we're left with.

One final observation concerns black and white portraits. As with people, this option is not necessarily a poor second to color, and may even be preferable to emphasize form, character and detail. While having no choice with my A20A and Rumpler CIV, since both now require restoration, I was not displeased to run Joe Faust's very professional black and whites. Neither subject is very colorful, in any case. For model detail and technique shots, black and white is usually preferable.

Also from the publisher

GERMAN FIGHTERS IN WORLD WAR II

THE DAY FIGHTERS

A Photographic History
of the German Tagjäger
1934-1945

Werner Held

THE NIGHT FIGHTERS

A Photographic History
of the German Nachtjäger
1940-1945

Werner Held/Holger Nauroth

Size: 7 3/4" x 10 1/2"
Hard cover, 224 pages, over 500 photos
ISBN: 0-88740-355-7 $29.95

Size: 7 3/4" x 10 1/2"
Hard cover, 232 pages, over 500 photos
ISBN: 0-88740-356-5 $29.95

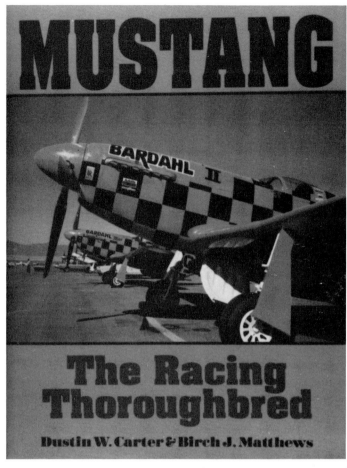

P-51 MUSTANG
A Photo Chronicle
Larry Davis

MUSTANG:
The Racing Thoroughbred
Dustin W. Carter/Birch J. Matthews

Size: 8 1/2" x 11"
Soft cover, 112 pages, over 200 photos
ISBN: 0-88740-411-1 $19.95

Size: 8 1/2" x 11"
Hard cover, 208 pages, over 180 photos
ISBN: 0-88740-391-3 $39.95

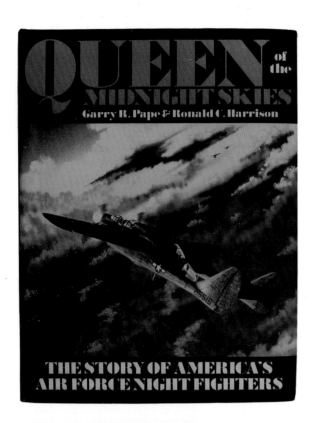

QUEEN of the MIDNIGHT SKIES

The Story of America's Air Force Night Fighters
Garry R. Pape & Ronald C. Harrison

Size: 8 1/2" x 11"
Hard cover, 368 pages, over 700 photos
ISBN: 0-88740-415-4 $45.00

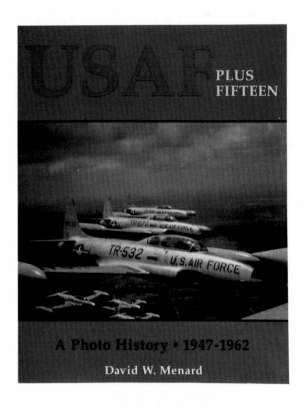

USAF PLUS FIFTEEN

A Photo Chronicle
David W. Menard

Size: 8 1/2" x 11"
Soft cover, 144 pages, over 400 photos
ISBN: 0-88740-483-9 $24.95

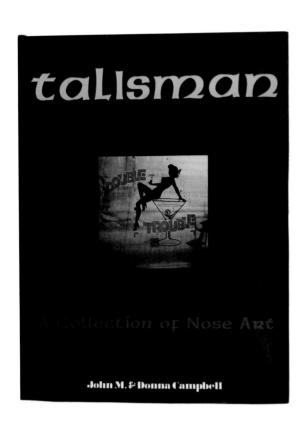

TALISMAN

A Collection of Nose Art
John M. & Donna Campbell

JAPANESE AIRCRAFT

Code Names & Designations
Robert C. Mikesh

Size: 9" x 12"
Hard cover, 256 pages, over 650 photos
ISBN: 0-88740-414-6 $49.95

Size: 6" x 9"
Soft cover, 192 pages, over 170 photos
ISBN: 0-88740-447-2 $14.95

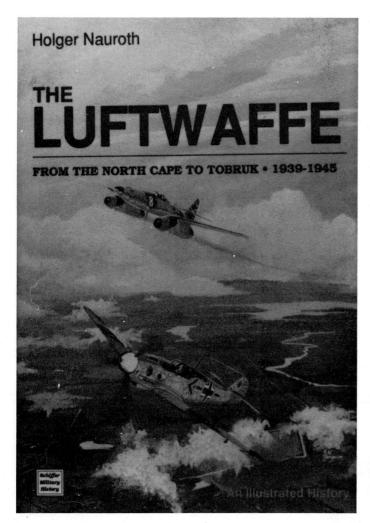

THE LUFTWAFFE

**From the North Cape to Tobruk
1939-1945 • An Illustrated History**

Holger Nauroth

Size: 7" x 10"
Hard cover, 240 pages, over 500 photos
ISBN: 0-88740-361-1 $29.95

THE LUFTWAFFE OVER NORTH AFRICA

**A Photo Chronicle
1941-1943**

Werner Held/Ernst Obermaier

Size: 7 3/4" x 10 1/2"
Hard cover, 238 pages, over 500 photos
ISBN: 0-88740-343-3 $29.95